Also by Tim Parks

Fiction

Nonfiction

Out of My Head

On the Trail of Consciousness

TIM PARKS

 New York Review Books New York

This is a New York Review Book
published by The New York Review of Books
435 Hudson Street, New York, NY 10014
www.nyrb.com

A CATALOG RECORD FOR THIS BOOK IS AVAILABLE FROM THE LIBRARY
OF CONGRESS.

ISBN 978-1-68137-397-3
Available as an electronic book; 978-1-68137-398-0

Printed in the United States of America on acid-free paper

10 9 8 7 6 5 4 3 2 1

To Riccardo
and Eleonora

Acknowledgements

I would like to offer my most sincere thanks to Jakob Köllhofer and Jutta Wagner at the Deutsch-Amerikanisches Institut in Heidelberg for their generosity in inviting me to the city in 2015 and making it possible for me to meet the scientists and thinkers whose work has enriched this book. Thanks also to those thinkers themselves – Sabina Pauen, Thomas Fuchs and Hannah Monyer – for giving me their time, sharing their ideas, having patience with my ignorance, but above all for the gift of their intellectual vitality, which was so encouraging and stimulating. I am very much in their debt.

Intellect to Senses: Ostensibly there is colour, ostensibly sweetness, ostensibly bitterness, actually only atoms and the void.

Senses to Intellect: Poor intellect, do you hope to defeat us while from us you borrow your evidence? Your victory is your defeat.

Democritus, fourth century BC

My old familiar opinions keep coming back, and against my will they capture my belief. It is as though they had a right to a place in my belief-system as a result of long occupation and the law of custom.

René Descartes, 1641

I am what is around me.

Wallace Stevens, 1917

Waking

I open my eyes and there is the wall.

No, that's not right.

I open my eyes and there *are* the wall, the wardrobe, the bedside table, the lamp, the tissues, the sheets, the blankets, the smell, the person next to me, the sound of the alarm. Multiplicity. I can't have one item without the rest.

But that's not quite right either.

I open my eyes and there are a *part* of the wardrobe – the side nearest to me, with a grey satiny wood surface – and a few surrounding patches of the wall which has a silver grey wallpaper with some stains around the bedside table, per- haps splashes of tea. The sheets are glimpsed, parts of the sheets, but also felt; the blanket has its weight, or rather mass, which I perceive as weight, thanks to gravity. I know the person beside me by the warmth and the breathing, but I haven't seen her yet. Also there's a window, though that's behind me surely. Yet I'm aware of it, or think I am, without seeing or touching it. I mean, I know it's there. I think I know. It's the light through the window, surely, that I'm seeing on the wardrobe and the wall. What else?

I close my eyes. Now the smell comes to the fore. What is it? Me, my partner, the room, the sheets, the carpet. It's warm. Or the breath making the smell is warm. Or my body. There's a strong feeling of my body that I wouldn't know how to describe at all. Eyes closed, waiting for the alarm to sound again, it is not exactly dark but not exactly light. More a kind of waiting to be dark or light when I open my eyes. For the moment I'm not seeing anything. But I'm not seeing nothing either. Perhaps I'm seeing the inside of my eyelids.

Would they be inside my head or outside?

My partner says, '*Amore*,' in a sleepy voice. And she asks, 'Are you cold?' I say no I'm not cold. If anything I'm hot. She's cold, she says.

I can feel a tug of bedclothes on my body. That makes sense: partner pulling the bedclothes. She has an issue with bedclothes. Awareness of my and my partner's history. Jokes about bedclothes. I could say something, but don't.

Suddenly I'm walking along a road by the edge of a wood. I turn to go in between the trees and see a stream at the bottom of a shallow valley, it seems a good place to swim . . .

The alarm sounds again. It's set for ten-minute intervals. I must have fallen asleep. So that was a dream and this is reality. The wood, the stream. In the dream I didn't know it was a dream, but nor would I have been able to say what came before the wood and the stream; I had no memory of the bedroom and the alarm; in the dream I was really in the moment; but now, back with the bedroom and the jingling alarm, I have a memory, or simply awareness, of the wood and the stream and there is some kind of continuity, a sort of me-ness that links them. In the one situation I can compare the experiences, in the other, I can't. Is that how I know this is reality and that is dream?

In any event, I feel I do know.

Again I open my eyes and see a whole that is made up of bits of all the separate things I see, none of which I see whole, as it were. I mean I see bits of the wardrobe and I could try to imagine the whole wardrobe as something you might walk round in an IKEA showroom, or I can imagine a two-dimensional photo or a drawing of the wardrobe, taken or done in such a way as to suggest three dimensions – I could even draw such a thing myself, come to think of it – but at the same time there are large areas of the wardrobe I will never see, where it backs onto the wall, for example, or underneath where it touches the floor. So when I say I see the wardrobe I mean I see that bit of it that is towards my eye and not blocked by the bedclothes. In fact when I use all these words, bedclothes, wardrobe, wall, lamp, I mean I see only the part of them that I see, though the word seems to refer to the whole thing, the idea of the whole thing. Words are Platonic maybe. They permit Platonism. Word 'wardrobe' = idea of wardrobe, not the bit of wardrobe I actually see.

Language is tricky.

I close my eyes but can't conjure up any clear image of an entire Platonic wardrobe, or absolute wardrobe, free from any contact with walls and floors. It all seems a lot of effort.

The alarm sounds *again*. These ten minutes went faster than the first.

Does that make sense?

When the alarm sounds my eyes open automatically. I don't seem to have much choice. It must be the famous conditioned reflex. Waking, you open your eyes and that's that. Once again I see bits of lots of things – or the parts of those things that are facing me – and now I realise *there are no gaps between these parts.* I mean there are probably scores of things I see (part of) if I bothered to list them – switch lampshade picture sock (one) tissues (many) carpet floor book wristwatch hinge of wardrobe door – but although I perceive these things as separate there are no spaces between them the way there are spaces between the words that designate them, even when I don't put commas in the list. The parts I see in the room run directly into the other parts I see. The world has no empty bits (that I can see). Or even fault lines like a jigsaw. Not even around the edges. It's seamless.

In all directions, then, as I turn my head, or move my eyes, the world continues and will continue wherever I go without any gaps. And the line at the peripheral horizon of vision is not sharp, as in a camera frame, but nor could you call it a fading out. I don't know how to describe it. There's no analogy. *It is as we know it.* There's an intensity of focus in the centre, the things I'm actively looking at, and then the sense of a periphery that could become focused if I so desired. But for the moment it's not focused. There's no edge there, whether hard or fuzzy, but as it were a nothing that might become a something all around the something I'm attending to that will become a nothing as I turn elsewhere. There's a constant feeling of potential, an invitation to move and change the things I am seeing, to substitute one thing for another. We rarely stay focused in one place for long.

'When do we have to get up?' my partner asks.

'We can wait one more alarm.'

Our hands are touching now. I can feel her fingers. She said she felt cold, and I feel hot but, touching, her hand feels hotter than mine. It's a pleasure to touch her.

Pleasure.

I close my eyes and try to remember some more of the dream, the wood, the stream. There is something there to remember, I think; something is telling me this dream was longer and richer than just the road, the wood, the stream.

But how can something tell me something when it's all me? Let's say I have a feeling that there is more to the dream than was immediately recalled. The road was rising slightly as I walked along it – a dark asphalt road – and the wood was to my left when I turned and entered it and saw between the tree trunks the grey earth sloping gently down to the stream. There was more. The stream was shallow. The water was transparent. Which was why I thought it would be nice to swim. I try to focus on this, or put pressure on it, whatever it is one does, but this causes an unpleasant feeling in my head. It's an effort that something inside me doesn't want me to make. There is tension. Between parts of me.

Leave be.

The room isn't there now I have my eyes closed, but nor are the wood and the stream. The smells and warmth are there of course, and the presence of my own body – I can't cut them out so easily – the same smells as when I had my eyes open, the same feeling of the sheets and the pleasure of the hand touching mine – I haven't seen it but I couldn't mistake it as anything but a hand – and the awareness of my partner who must be there otherwise the hand wouldn't be there, and her voice. A faint sound of traffic too.

Is the road outside *wet*? I'm not sure.

Eyes closed, the room isn't there, but I know I just saw the room and quite recently the wood and the stream. Saying I know I just saw the room is not the same as saying I remember the room. Where exactly is that sock, for example? Blue sock or brown? Don't know. I remember the sock but not its colour. Actually I don't even remember the colour of the carpet, though I know this room has one.

So does my experience of the room have the same status as my experience of the wood and the stream? I'm not sure, I'm not sure what status means really. Was it as *real*? Is that what I mean? Everything is real while it's happening, isn't it? Or it wouldn't be happening. That is: something must be happening even if it might not be the thing you think it is. Otherwise there would be no experience.

Feeling my partner's fingers stroking my wrist I'm aware of an enormous reluctance to get up; being in bed is so nice and it's a damp autumn day. How do I know it's damp if I haven't looked out of the window and couldn't tell whether the traffic sound suggested a wet road or not? I don't know, but I feel as if I did.

The truth is, today will be busy. Not that my schedule is immediately present to me in all its details, but I'm aware that pretty soon I'll be checking it on my computer and doing all the things I've arranged to do, keeping the appointments I've set up and calling the people I need to call to set up further appointments. In short, I'll be making a big effort to be the person I am very aware of being as I wake up.

But who is that person?

The person who made these appointments of course. Who would I be if I just dropped my appointments and stayed in bed? Who would pay for the hotel room?

My body feels so cosy in the sheets in the dark brightness of closed eyes and then I remember there were animals in the water. I turned left off the road, walked into the trees and saw a broad stream with animals all walking upstream from left to right. A leopard. A dog. A very large white rabbit. Largest of all, some kind of dinosaur, a man-size lizard walking on hind legs, kind of thing. But without a head, or not something you could easily call a head. I remember thinking this was strange. Like a toy, but life-size.

Do I see these animals again now?

Not really. I'm aware of having seen them. I know I had that experience. I think large white rabbit. Using words. I saw a large white rabbit. It was hopping in a standing position, not running the way rabbits really do. It was very white. Like the white rabbits I kept as a child. But I wouldn't say I have a clear image of the rabbit in my head, like a photo, or a video. On the other hand I know I'm not just using the word. It feels there is something visual there, a sort of waiting for a rabbit to reappear, or the after-image of a sudden rabbit when it's gone, but not something I can really see, the way I would immediately see the hotel room if I opened my eyes, which I don't want to, snuggled as I am in the sheets with my partner stroking my hand, feeling extremely comfortable and not wanting to think about today and everything I'm supposed to do. It will be busy.

Very soon the alarm will surely go off again. It's been quite a long time already.

Why did I dream of animals? There was an emotion of surprise in the dream when the animals appeared, perhaps because they were oddly sized and walking upstream. The rabbit was bigger than it should have been. The leopard and the dog were smaller and nondescript. The dog is only a word now, there is nothing left of it visually, though there must have been something initially, otherwise why would I have thought, dog? Still, I thought leopard too and I'm not really sure I could recognise a leopard if I saw one, I mean, distinguish it from, say, a cheetah, or a jaguar. In any event there was something catlike that I took to be a leopard. Or perhaps I dreamed the word, leopard, associating it with the catlike creature I was seeing. Going back now, to the feeling of surprise on seeing the animals marching together up the shallow stream, it wasn't unpleasant at all; it was a mildly hallucinatory, dazed sensation, as if something significant were being revealed; and the significance, whatever it was, corresponded to this sensation, this dream state, which of course ended abruptly when the alarm went off and I opened my eyes to the room.

Is there a gap between the wood and stream and the room?

Yes and no. The wood isn't anywhere now, it's not next door to the room, but I am the same person who saw it one minute then the room the next.

Aren't I?

Or could I have changed in the waking? Are such things possible?

8

I *feel* I am the same person. I feel I give continuity, even if only a posteriori, to the two experiences, dream and reality, the same way I feel I am the same person I was forty years ago, though meeting me forty years ago must have been pretty different from meeting me now.

This is shaky ground.

What we can say is that whatever the nature of the continuity, it's quite different from the continuity that words give. I can control the continuity of the words (I believe), their being strung together in sentences and paragraphs and essays and stories, but not the continuity of seeing and feeling. Seeing and feeling, everything fits together and there are no empty spaces as my experience moves from the hand stroking my arm to the stream with the headless dinosaur and the white rabbit, then the traffic noises beyond the window, and back to the stain on the wall beside the lamp with the green shade.

Experience is more continuous and more *happening at once*, as it were, than words where only one thing can be said at a time. Words are linear. They sort things out, pull them out of the mix, conjure each object whole, when I don't see anything whole, ever.

None of us have ever in our lives seen a single object whole, from all sides, up and down, the way a word denotes it.

But then again, words themselves are only part of the experience mix; there is no space between words and sensations; my partner, for example, now whispering something sweet in my neck as I begin to tense myself for the sound of the alarm which must surely come at any moment.

It's a little singsong. The alarm. Der, d d d der der!

Actually it's my partner's mobile. Words on the page can't do the singsong. Nothing I can say or write, however technical or poetic, can give to you, dear reader, the hateful sound of this alarm which goes off every morning at 7.30, for about ten seconds, Der, d d d der der, then again at ten-minute intervals until at last we get up and turn it off. Hopefully before eight. Of course it is only hateful because it wakes us up. Otherwise it's not an unpleasant sound at all. It's the same sound the phone makes when you turn it on. Perhaps I could write something witty and evocative about the sound of the alarm, if I felt inspired. Maybe you would appreciate that. Maybe then you would say: the writer really made me hear the tinny sound of the alarm, and how hateful it was. But you'd be kidding yourself. The witty, evocative description would only give you something witty and evocative. A piece of writing. Words. Not really the sound of the alarm which is going to happen any second now. In fact it seems amazing it hasn't gone off yet. Has the battery died?

I keep my eyes closed, knowing that when I open them in response to the alarm the world will be there and now it really will be time to get going. Aside from everything else, I'm aware of needing a pee. Actually, now I tell myself I need a pee I'm aware of having been aware of this for a while though I didn't mention it to myself, as it were. How many selves do I have? Perhaps I mean I didn't *focus* on it. The focusing business, that is in the continuous me, didn't focus on the need to pee. Until now. Needing to pee was there, waiting to be brought into focus. Surprisingly patient.

What if I opened my eyes now *and the world were not there?*

Is such a thing possible?

It would mean I had gone blind, presumably. It couldn't mean the world really wasn't there because I would still be able to touch and feel things. If nothing else, I would feel the support of the bed beneath me, and of course the smells. I would just be blind.

But what if I woke and *there were no smells either?* And no sense of the bed supporting me from beneath. At that point I'd be a vegetable. I'd be brain dead.

But does a vegetable wake up?

What if I felt convinced that I still had all my faculties, that I was perfectly awake and capable of seeing, hearing, smelling, touching, etc., *just that there was nothing there for me to see hear smell touch?*

How weird would that be? A sort of floating free. Ultimate detachment. I'm not sure I can imagine it even.

The reality is that when the alarm goes off – it's been an amazingly long time now – and I open my eyes, the world will be there just as surely as I am there and my body is there. In a way you could say the world is as reliably attached to me as my body. Or I to it. I mean, I can move my eyes to different parts of the room, the world, but *I can't separate seeing and the world.* We're a whole.

Perhaps it's like when, as a child, you pushed a suction cup against a wet windowpane and couldn't pull it off. You could move the rubber cup across the pane, maybe it was a plastic arrow you had fired at the window, but you couldn't detach it.

What a crazy analogy. Where did it come from? Was it stored in the brain somewhere, or instantly concocted?

The fact is I can feel, or at least remember, feel-remember, my hand tugging gently on the toy arrow and the suction cup resisting, sliding across the glass, but refusing to come free; a bizarre collaboration between childhood experience, brain chemistry and language. We're back with words again. I wouldn't have come up with this analogy if it wasn't for language, which loves analogies. In fact, the only way I can maybe detach myself from the world, or enjoy that illusion, I mean detach myself from the immediate world of the room, of this morning, here in Heidelberg, September 2015, is by concentrating on the words and a sort of momentum they have when they get going so that the things in the solid world around you slip into the background. Because my attention isn't focusing on them, but on the words.

In that case, then, to separate myself from the world I need to keep talking to myself, ten to the dozen. However on this occasion my talking to myself has led to the conclusion that I can't ever really detach myself, my waking consciousness, from the world around me. When I open my eyes I just won't be able not to see the silver grey wallpaper. We're suckered together.

Could you say then that the world immediately around me is *as much me as my body*? In the sense that it's equally *inalienable*?

And are words too perhaps just another manifestation of the world? Another category of things I couldn't shake off even if I wanted to?

In which case I really am glued to experience day in day out.

'Ele?'

'What?'

'Don't you think it's been more than ten minutes?'

'What?'

'The alarm. Do you think the battery's dead maybe? It seems it should have gone off by now.'

'Just enjoy it.'

Der d d d der der. The alarm sounds as she speaks.

I open my eyes and there's the wall. The wallpaper. Inalienable.

You go first, my partner says. The sock is blue.

Colours

Let us pass over some experiences on the way to break-fast. I'm in Heidelberg to meet various professors and talk about what consciousness is. I'm in Heidelberg to spend a few pleasant days with my partner. Stepping into the hotel breakfast room it's hard not to be aware of possible eyes on us. My partner is half my age. Slightly less to tell the truth; the tipping point will come in a year or so when she hits thirty-one. Figure it out for yourself. Such asymmetrical relationships are supposed to have something wrong with them. Various authorities, religious and secular, maintain that it's far healthier when couples are more or less the same age. This view does not correspond to our experience, mine and my partner's I mean; to us our relationship seems absolutely fine. In fact, it's rather shocking to discover how wrong the alleged authorities can be in this respect. How many other things might they be mistaken about, you wonder? For example, my friend Riccardo Manzotti has recently demonstrated that the current model purporting to explain what happens when we experience colour after-images is quite wrong. But that is another conversation.

Right now, taking a table in the breakfast room, we're aware of two or three gazes assessing whether we are father and daughter or older man and mistress enjoying an illicit fling. But how could we be father and daughter, when we look so completely different? People sense these things at once; vast quantities of experience are brought to bear without even thinking, I mean without consciously reflecting. The sharp observer simply *knows*. The hotel maid, for example, a big girl in her late teens trussed up in a starched white apron with a little black cap, immediately assumes an air of complicity; she's going to be well disposed to our supposed transgression.

And this would be the kind of thing one could so easily write about of course: how two people's lives suddenly tangle together in the most unexpected and incongruous ways and how other people react to that. That is the stuff books are made of. Usually. I mean novels. Because that's the way we tend to think of our lives, projecting things backward and forward in stories. And also because words, sentences, which run linear fashion through time, gathering energy as they go, are good at that, good at telling stories that also move through time: people meeting and falling in and out of love, finding and losing jobs, striving and succeeding, striving and failing. In fact, a lot of our lives seem to be made up of words, imposing some kind of shape and momentum on the precarious, barely describable business of actually being here, moment by moment, in the world: waking up to the wardrobe and the wallpaper, drifting in and out of dreams while the ten-minute interval of the alarm seems first cruelly brief, then disturbingly long. *Stories are more familiar to us than living itself.* Easier to be inside. Which is why we like novels, I suppose. Biographies, histories and memoirs too. I love them. Experience unfolds in them in a way that makes sense, altogether ignoring those redundant experiences between waking and breakfast: the obligatory visits to the bathroom, the struggle to figure out how the hotel shower works, the dazzled feel moving among gleaming reflective surfaces, the need to put your glasses back on to read the small print on the tiny coloured bottles: shampoo or shower soap? How could you ever keep a story going, or even an argument about the nature of consciousness for that matter, if you started to look at how life, how consciousness, actually *is*?

No, if we start writing about the present moment and all the whirl of perception and cogitation that goes with it, we'll be overwhelmed. We'll never capture it all. And we'll bore the reader silly in the process. The genius of language is omission. It misses most things out, almost everything in fact; it invites the reader to board a fast train across the unnecessarily cluttered landscape of ordinary experience.

Or if we do try to say it all, to describe what being alive is like, I mean, at any specific moment, we'll be so worried about boring our readers that we'll start to make it fancy, start to redeem the avalanche of moment-by-moment perceptions in a little rhythm and rhyme, a little poetry maybe. We'll try to make our account *attractive. Ulysses*-like. But what will be attractive will be the writing, not the moment itself. We'll start getting interested in the delivery rather than the thing delivered, the way everyone is far more interested in *Ulysses* the book, its stream-of-consciousness style and controversial author James Joyce, than in anything his book actually talks about: Dublin, masturbation, newspaper advertising, funerals, prostitutes.

I want to avoid that. In fact the problem that faces me on this trip to Heidelberg is how to focus on the business of being conscious, how to invite *you* to focus on it, but *without literary intent.* We're not going to do prettiness. Or melodrama. Or even polemics. We're simply going to ask: can we ordinary folks say something useful about consciousness, observing our own experience, from moment to moment, these next couple of days? And then we're going to ask, do the models, the explanations, whatever, that we have of consciousness, the version of events that our various authorities sign up to, make sense? Do they fit with what we experience second by second? And if not why not? Fortunately we've booked appointments with some smart people to discuss this.

All the same, I can't help stopping to consider the glories of the breakfast buffet in this Heidelberg hotel. Basically there's a black stone counter around two walls in a corner. At least it looks like stone. Then there are silver trays on this counter laden with fruits, pale green melons cut into slices, pineapples opened in symmetrical triangles, plump round strawberries laid one by one directly on the silver tray, figs slit open to display their meaty interior (just slightly more maroony than the shiny red of the strawberries), and in the middle of the tray, I mean of each tray, the bristling flourish of a pineapple top, its circles of dark green leaves sprouting up one inside the other then folding outward in a triumphant, crest-like flourish of vegetation.

Between the trays – sorry, we haven't finished yet – are three-tiered bowls raising apples and grapes and oranges and kiwis high above the other foods, and candles in silver candlesticks with electric lights on top and, because this is Germany, plates with gherkins and radishes split open to enjoy the contrast of white inside and pink out, and thin slices of tomatoes and red, green and orange peppers, arranged symmetrically on beds of lettuce, with slices of cheese and rolls of ham beside bowls of boiled eggs and baskets of crusty bread rolls with sesame seeds and dark brown loaves wrapped tight in white cloths so you can slice them without transferring your bacteria to the bread.

Behind this cornucopia, along both walls of the corner where the counter is, runs a mirror just high enough to double this explosion of colour which is made all the more intense by bright overhead spotlights sunk into the ceiling. Actually, more than double it, since from the angle I'm standing at to fork fruit into my bowl, the mirror in front of me gives back not just the fruit, but the mirror on the other wall too, in a dazzling multiplication of wonderful things to eat wonderfully arranged and constantly replenished by the attentive hotel staff.

All this without even mentioning the gleaming white plates and polished cutlery and the tall silver urn with boiling water and above that a shelf of teapots and of course the inevitable Twinings tea dispenser with its prettily coloured sachets. The book I have just been reading, *The Quest for Consciousness* by Christof Koch, describes colour and indeed perception in general as a 'con job': there are no blues or greens or reds out there in the world, Koch says, they are all generated in your brain, a cooperation between the visual cortex and three groups of clever cone-shaped cells in the retina transforming the light frequencies bouncing off different surfaces into the illusion of colour. In reality, then, and Koch is pretty much in line with received opinion here – 'received opinion' and 'various authorities' being more or less equivalent terms – all this breakfast buffet is happening in my head and the strawberry I just spooned into my bowl and will soon pop in my mouth is red nowhere but in my head. Although of course my brain is famously made up of grey gelatinous stuff.

Shaking my head (a kaleidoscope of colours), I try to figure out the tea urn. It's a handsome thing, quite tall with a tap at the bottom and a kettle affair on top. Near the tap is a red switch, but no indication of whether it's on or off. So is the water hot or isn't it? Can I find out by touching the outside of the urn, or is the insulation such that it always remains cool? Am I supposed to turn the thing on myself? Assuming it's presently turned off. Is that permitted? Or should I ask the waitress? She has disappeared. I hate tea that is not hot. And do I remove the kettle from the top and pour from that, or do I put my cup, or one of the other smaller teapots under the tap beneath the urn? I don't know. The hotel staff are obviously under strict orders to keep everything very shiny because as I bend to study the urn I see my face in its silvery finish, though distorted of course, stretched horizontally as the urn curves away from me. The con job of perception plus the distortion of the curved mirror. Yet I recognise clearly enough the puzzled look on my face as I choose the tap and get it wrong, my tea is barely warm, damn!

Back at the table, we debate whether I should have another attempt at the urn. My partner thinks I should ask the maid, but neither of us speaks German, or not fluently, and I don't want to appear stupid. Why not? Who cares whether the maid thinks I am stupid? She is not an authority. Or even a peer. And she already seems to think I'm here at the hotel with my bit on the side. But apparently I *do* care. The maid is a human being and I don't want other human beings to judge that I don't know what's what when it comes to tea urns.

Or is it rather that I want the pleasure of figuring out the urn for myself? I want the small triumph of securing myself my hot water all on my own. The woman I am going to see this morning, Professor Sabina Pauen, sets up experiments about children's ability to learn at very slightly different ages. Can a child at twenty-two months distinguish between functional and non-functional features of a simple tool? What about at twenty-four months? 'Typical experiment' – I explain to my partner as she checks the nutritional info on a yoghurt carton – 'there is a long transparent plastic tube with a reward stuck in it. To push the reward out you need a stick-like instrument long enough to go right through the tube. There are three instruments of different lengths with handles of different colours. Only one is long enough to do the job. The kid sees an adult take one of the sticks, with, say, a blue handle, push the reward out and make a delighted face, after which the reward is given to the child. Then the child is handed a new tube, reward inside, together with the three tools. Does the child understand what to do? Does he or she choose the right tool, the one with the blue handle?'

'What's the reward?' my partner asks.

'No idea. Not a hot cup of tea, obviously.'

'It might be important to know.'

'Anyway, when the kid has figured it out, they take a break, then repeat the experiment, but changing the handles on the tools so that the tool that works now has, say, a red handle. The blue won't work this time.'

'Cruel.'

'So the question is: can a toddler distinguish between the functionally relevant length and the functionally irrelevant colour?'

'What if they don't really want the reward?'

'Maybe they'd be eager to impress their parents anyway, or just themselves. They want to feel good, that they figured it out. Regardless of the reward.'

'Are the parents there?'

'One parent.'

'Paid?'

'It seems not.'

'So we have parents interested in child development.'

'In achievement, yes. No doubt they want their kids to get on.'

'But what if the child rebels against the parents' achievement obsession and refuses to get interested in the tube game?'

I shake my head. 'The fact is a lot of these experiments are done with electrodes stuck to the kids' heads so they can register brain activity in the frontal cortex.'

'Go and get your tea,' my partner says.

'But what am I supposed to do with the cup I have? It is still full.'

'Leave it and get another.'

I'm reluctant to do this, thinking of the maid shaking her head over my wastefulness as she clears the table.

'I hate to seem wasteful,' I say.

'You're ridiculous,' my partner tells me. 'Go.'

I cheat. I wait for someone else to go to the urn and watch what he does. It's a middle-aged German man. He presses the red button on the side of the urn which now lights up. It hadn't seemed like the kind with a light inside. I'm incompetent. It was easy. He waits. I go and wait behind him. There are about a dozen people in the breakfast room which has tall, straight-backed chairs, all a silvery grey (in my head), a fitted carpet, which (in my head) is black with small white crowns in a pattern running diagonally to the rectangle of the room. What I would like to ask Christof Koch is whether the colours are more of a con job than the shape and size of things, especially since Koch is convinced that our entire experience of the world is located in the head, an image, as it were, or representation, in the theatre of the mind. Galileo of course thought that the only things really out there in the world were shapes, numbers and movements; everything else – colour, sound, taste, smell, etc. – was just added in by eyes, ears, tongues and noses. Merely subjective, not even existing outside the human race. So that really to know an object you had to measure it in every possible way, with scientific instruments, and study its shape and mass and movements, but none of the things that might normally engage you, its colour, smell, feel. Perhaps this is why I never took up science.

My tea comes out piping hot. I now know something

more about tea urns. Of a certain kind. Never again will I be humiliated by lukewarm tea in a German hotel breakfast room. So presumably something has shifted in my brain. It is different from what it was ten minutes ago. This is part of the burden of Professor Pauen's research, that the child's brain, in its period of maximum plasticity, is actually *changed* by the little learning games her team takes him or her through. In one experiment, for example, young children are shown sequences of photos very rapidly, perhaps faces, both familiar and unfamiliar, male and female, perhaps animals, perhaps items of furniture. The child's response is measured through an electroencephalogram which records brainwaves – surges of electrical activity in the brain – their intensity and the exact number of milliseconds they occur after seeing the photo. Result: after scores of nine-month-old babies have been sat in front of a screen with electrodes attached to their heads it seems fairly certain that they can distinguish between categories of animals and furniture and between men and women, men's and women's faces that is. Rather more interestingly, it's also clear that they react less intensely when a photo is a repeat of a previous photo, even when they only saw that previous photo for half a second. So, in that very brief space of time the brain changed, it learned something, as I now have learned that I have to switch the urn on and wait before turning the tap for my hot water, even though I have not learned what the cute silver teapot on the top of the urn is about. Is it a functional or a non-functional feature?

Returning to my seat, second cup of Twinings Breakfast Tea in hand, the room is abruptly altered by an apparition. A woman sails in with a bright yellow hat.

I should say that all the other folk in the room could be described, at first glance, as nondescript. Is that a contradiction in terms? You know what I mean and I do not intend to be disparaging. I too am nondescript. None of these people are sending out any particular signals or inviting attention in any way. There are the middle-aged German man and his wife to the left of us, a young Japanese couple sitting by the window, an Arab family, husband, wife and remarkably well-behaved toddler. All dressed in the most ordinary ways. The Arab mother's black headscarf seems utterly ordinary. Aside from that she is wearing jacket and jeans.

Everyone is going about their breakfast business in a discreet fashion. Which means I suppose that we share a sense of how one behaves at breakfast in hotels. Any subdued conversation is rendered indistinguishable from table to table by softly played pop music of the 1970s variety, 'Super Trouper', for example, at this particular moment. Which again seems to fit with the casual clothes, the hush of the carpet, the grey (in our heads) upholstery. In fact, I hardly noticed any of this until now. We were all integrating in the hotel milieu, making ourselves invisible. Perhaps conforming is the word. Is this because the millions upon millions of similar impressions our brains have received over the years have rendered us somewhat similar, or at least similarly disposed (programmed, wired?) for having breakfast in hotels? All the scientists I read who are studying the brain use computing analogies. Pauen's babies 'process' images and 'encode' and 'store' them. It seems the natural way to talk about the brain, though of course there's nothing natural about computers. Or breakfast in hotels.

In any event, across the unrippled surface of this early morning conformity, somebody tosses the pebble of a bright yellow hat. Actually, our new arrival is all yellow: yellow shoes, yellow blouse, yellow cardigan, yellow fingernails. But it's the hat that does it. And when I say bright, I mean *bright*. A hat at once voluptuously round, yet seemingly rigid. Flat on top, a sort of inverted saucepan pressed onto permed hair, but with a very generous brim, perhaps two feet in diameter. It seems to pull all the room's light towards itself and then send it bouncing off in a lemony dazzle. After her, and she moves with a very female sway, trails both a heady perfume and a rather less exciting husband. The perfume is flowery and fruity. The husband is dressed in white.

If colours and smells are a subjective con job, nevertheless this lady, in her early sixties I'd reckon, certainly knows how to predict and manipulate our response to chromatic effects. Her hat turns every head. She takes a table by the window, beams a charmingly complacent smile around the room as if to welcome us into her atmosphere of sunny self-congratulation – even her lipstick is yellow – then calls cheerfully to the young waitress, whom she knows by name. Her husband is admirably at ease with her exhibitionism. Even proud perhaps. He too greets the waitress and unfolds the *Frankfurter Allgemeine.*

'Of course, the yellow is all in our heads,' I remind my partner. 'And the smell.'

'Still, there's no getting away from it,' she points out. 'Even if we wanted to, we couldn't stop making it up. And we do both call it yellow.'

I had been meaning to use our breakfast chat to discuss

Professor Pauen's experiments and thus prepare myself for my interview. Instead it's impossible not to talk about this yellow. We've been taken over. The lady is elaborately made-up, yet exudes a sort of performed naturalness. And the performance enchants. Why is she so different from the rest of us? How come she and her husband can order food to be brought to their table while we had to go and forage for ours at the buffet, arriving at a decent cup of tea only through a process of trial and error? Is it that while our brains are all programmed, or resigned, to conformity, she, having received different impressions over the years, or, more likely, having processed (whatever that means) the same impressions very differently, is a free spirit? Or is she simply obeying different imperatives? Her behaviour is, as my partner remarks, using an Italian idiom, '*tutt'un programma*'. She didn't invent this look. It seems to have been cut out of the Kodachrome photography of a 1960s fashion magazine; it's something she once saw, I suggest, that sank so deep into her psyche that she is still reproducing it.

No. My partner doesn't see this lady as someone who has succumbed to an impression, rather as an agent who calculates the most effective strategy for drawing all attention to herself. Women are always the harshest critics of other women. In any event, I'm aware, folding my napkin, that the yellow hat has made such an impression on me that I will always remember this breakfast, this hotel, because of it. I will be able to start with the yellow hat and then retrieve the rest of the scene from that; my position in the room in relation to the wearer of the hat, and hence to all the others, the Japanese, the Arabs, the middle-aged German couple, the windows and the splendid spread of food mirrored from two walls in all its silver and fruity colours, much the way one remembers a dream starting with the one fragment you were left with on waking, the view of the stream through the tall trees of the wood.

But now it's time to go upstairs and prepare for my interview.

Inside Out

I now have forty minutes to look through my notes before the first interview. But let's recap. About two years ago I received an email from a man called Jakob Köllhofer inviting me to participate in a project which involved 'sending writers to the scientists of various departments of Heidelberg University to find out whether the "Sciences" could come up with a concept of "new metaphysics"'. I was intrigued by Köllhofer's use of inverted commas but perplexed by the idea. How can observation of the world and even consequent reflection and speculation ever lead one to the *why* of the world's existence? One might prove the Big Bang theory, but learn nothing of why it happened, or why it happened when it did. Or whether it *had to* happen. And even imagining one did learn all those things, there would be the enigma of what came before, and then before that again.

However, after a further exchange of mails it emerged that Köllhofer was primarily interested in the question, whether *in people's minds* science was substituting religion. This seemed more a matter of anthropology than hard science, something I might reasonably comment on. Also, the original invitation said that Köllhofer's employer, the Deutsch-Amerikanisches Institut, could offer 'a considerable honorarium'. Who would not be curious to know what was meant by 'considerable' in a phrase like this? And then there would be the opportunity to meet some top scientists and talk about consciousness.

Because for some years now I have been fascinated by consciousness, hungry and thirsty to think and talk about it. Which is as much as to say that for some years I have been in an intense conversation with Riccardo Manzotti, one of the most intense and extended conversations of my life. And in fact, returning to our hotel room and sitting at the tiny desk provided, I made the mistake of checking my email before looking at my notes on Sabina Pauen. Sure enough there was a message from Riccardo including a link to a review by the philosopher Alva Noë of the film *Inside Out*. It's not every day you see a philosopher reviewing a film, so I now made the further mistake of glancing through this review, rather than rereading Pauen's paper on 'Object Categorization and Socially Guided Object Learning in Infancy'.

No doubt you remember *Inside Out*, a Pixar animation. A girl, Riley, goes through some emotional turmoil when her father gets a job in a different town and she has to move and adapt. But the big idea of the movie is to portray Riley's psyche in the form of five tiny people dashing about inside her head, all competing for the controls that lead her to act and respond in this or that way. These little homunculi, uncannily similar to the characters in other Pixar animations, explicitly represent the emotions Joy, Sadness, Anger, Fear and Disgust, but since they also seem to be fully formed individuals the film raises the question, which Alva Noë claims his own son put to him halfway through the movie, as to whether these tiny folks inside Riley's brain have further and tinier folks inside their brains. Also, what is it that gives Riley the sense of being one person rather than five, or five times five, or five times five times five? And so on.

It's hardly surprising that a Pixar animation might not be the *non plus ultra* when it comes to psychology, ontology or metaphysics, but the interesting thing is that *Inside Out* was widely acclaimed as innovative, insightful and educative, this as it presented poor Riley as little more than a puppet at the whim of her five incompetent puppeteers. Is this really how most people think about the relationship of mind to body? Noë glosses:

> Descartes (1596–1650) offered, but did not endorse, the idea that the body is a ship and the self resides in the body the way a pilot resides in the ship. Hume (1711–1776) advanced the idea that there is no self, that what we call the self is in fact just a bundle of perceptions, feelings and ideas. Contemporary cognitive science combines these two ideas in a most awkward synthesis: we are the brain, which in turn is modeled not as a self, but as a vast army of little selves, or agencies, whose collective operations give rise to what looks, from the outside, like a single person.

Reading this I can't help but envy Noë's powers of synthesis, which are very far from awkward, and I wonder if Sabina Pauen has seen *Inside Out* and would be willing to discuss the issues it raises: above all, is conscious experience really all locked inside the head, with people inside people peering out at each other as if through periscopes (an effect achieved in *Inside Out* by having the homunculi use their eyes to look at screens which record the material captured by the cameras that are Riley's eyes). Or is it, conscious experience, in some way also *outside*, or at least actively composed of both inside and out, the person experiencing and the thing experienced? Because although I'm ostensibly here to write a paper for Herr Köllhofer in response to the question, is science substituting religion in people's minds? – that's what I'll be receiving the considerable sum of €10,000 for – my deeper desire is to hear what the scientists I'm going to meet think about this inside–outside question, which at bottom is nothing other than the nature of our engagement with the world and the deep puzzle of minute-by-minute perception: that what is out there when I open my eyes in the morning – the wall, the wardrobe, the bedclothes, my partner – is also as it were inside me, or at least present to me, in some way possessed by me, and hence in two places at once, the wardrobe in the room and the wardrobe in my head. How can this be and what is really happening? This is the question that fascinates me, the subject of all my conversations with Riccardo Manzotti, the object of interminable reading, and, in a way, the question one asks in every novel: how does the experience of each separate character integrate with the world as a whole, including the experience of everyone else? How separate are we, and how bound together? Islands entire onto ourselves, or pieces of the continent, parts of the main?

Am I here, then, in Heidelberg under false pretences, or at least grabbing my 'considerable sum' with my clumsy left hand, while my more competent (hopefully) right is up to something else? Not exactly. For nothing could be more important when considering the competing claims of science and religion than the question of whether the self, the mind, the soul, or just consciousness, is a separate thing, isolated in the head, or some ongoing collaboration between body, brain and world. If the former – the mind separate and isolated – then we have the dream of abstraction, the possibility that the thing I think of as 'I' might in some way be salvageable from the ultimate and inevitable wreck of the body, might even fly off to heaven perhaps, as the Christians would have it, or simply be uploaded into some extraordinary computer, as a number of neuroscientists now claim will one day be possible. If on the contrary the mind is a phenomenon that has to do with the body's interaction with the world, or the world's with the body, if indeed it is sustained by and dependent on that interaction, then it will simply stop when the body dies.

Or again, if the former – the mind marooned in the head, pulling levers and pushing buttons (alone or in a team) to tell the body what to do – then our knowledge of the outside world will always be suspect. How can I know a world if I'm not part of it, if I'm stuck in Plato's cave unable to experience the reality without, if I'm seeing colours where there are no colours, smelling smells when, as Galileo would have it, there are no smells? I might be Paradise-bound or potentially immortal in microchips, but my experience will all be, as Christof Koch tells me, 'a con job'. I'm an anomaly on a planet quite different from the one I perceive, making stuff up that just isn't there. Always doubt your senses, Bacon tells us. It's disquieting.

But if the latter – the mind a consequence of the meeting of body and world – then the mind is actually one with reality, for at that point the mind is the happening of body and environment together, the environment including, of course, other people; so, far from being isolated and deluded, in this scenario the mind becomes the proof of a real encounter. In which case I might not be immortal, but equally I won't be needing a priest or neuroscientist to tell me what is what. My experience *is* what is what. It's really happening.

So a lot rides on Alva Noë's scathing review of *Inside Out*, which, scrolling down the screen with the aid of the Hotel Panorama's excellent Wi-Fi, I see was loathed by most of those who commented on it, people who generally found the film charming, delightful, and simply can't see the philosopher's problem; he's a boffin, a bore, they complain, though in fact, reacting to the name Alva, most of them assume that Noë is a she not a he. Their minds are swayed by cultural norms; foreign names ending with an 'a' will usually be feminine. 'She needs help!' somebody protests. And this of course, it occurs to me, is what I too will be up against if I suggest that this now standard view of conscious experience as something locked away in the head is nonsense. The vast majority of people seem quite comfortable with the idea. It has the status of a cultural norm, like the language we speak, or our belief in democracy, or simply good manners. Going against it will not be easy.

But time is getting away and good manners demand that

I not be late for my 10.30 appointment. According to Google maps Sabina Pauen's office is just five minutes' walk from here, so I'm going to give myself ten. I thus have twenty to look through my notes.

Essentially, Pauen has sent me three research papers, one about face recognition in babies of nine months old, one about how toddlers acquire knowledge about tools, one seeking to measure how one-year-olds respond differently if their reaction to a person or object is guided by an adult. Some of the terminology is daunting. The face-recognition experiment seems simple initially, nothing more than babies being shown photographs, but it becomes immensely complex when one looks at the details. The babies have to be the same age, the same race, the same class, more or less, but divided equally into the two sexes. The photos have to be standardised and vetted, all Caucasian faces of similar shape and size, neutral expressions, similar luminescence and colouring.

The sequencing is crucial. Are the babies to be 'primed', that is shown an image before the recognition trial proper begins, this to separate the trial experience from the experience that went before? If so, is that prime itself to be a face, or some other object, or a geometric design? Are the babies to be shown an equal number of men and women? Of what ages? Alternately or at random? Are extremely ugly or extremely beautiful faces, faces that might distract from the focus on male/female, to be included or not? How long is each photo to be shown for? A few milliseconds, a second, two seconds? How often are faces to repeat, and is repetition to be regularly sequenced or random? Etc. Etc.

All the babies have to be sat before the same-sized screen,

at the same distance, and on the mother's lap, but the mother must be careful not to interact with the baby. How easy is that? As the test proceeds someone has to check whether the baby is actually looking at the screen and, if not, this has to be recorded. Meantime the baby has a sort of hairnet studded with electrodes put over its head so that an encephalogram can record the various electronic waves beneath the surface of its skull. Attention will be focused on wavelengths associated with face recognition, which, incidentally, are not the same for children as for adults. The encephalogram is synchronised with the slide show so that afterwards it will be easy to determine the response to each picture. Where the child wasn't paying attention the result is ignored. Or sometimes another event occurs, a hiccup, an itch, in which case, again, the result has to be discarded. Some babies pay so little attention that even the few moments when they are doing can't be taken into account, for how can the whole effect of the sequence, the repetitions and so on, be considered if the baby hasn't looked at very much? Results from 60% of babies tested are discarded.

After the trials themselves are over, the real work begins. Which wavelengths occurred? Exactly how long after each photo was shown? With what amplitude? Is the response different when the child has been primed? Is the response to male and female faces the same? Is the response the same when a face is repeated? Is this alteration in response to repeated faces the same for men's faces and women's faces? Does this change depending on whether the child is male or female?

To get my mind round all of this I have had to read up on

brain geography and encephalography – what exactly can the encephalogram tell us, what are its limitations – and then to remind myself of some basic statistical computations when it comes to recording results. All this work has been taxing, but fascinating. The maniacal attention to detail and the precariousness of the actual laboratory test – simply getting a baby to pay attention, getting a mother not to worry about those electrodes on the child's head – is extraordinary. And what if the person checking that the baby's eyes are watching the photos should themselves experience a moment's distraction?

This is science. The dogged search for controlled, comparable, repeatable conditions so that something can be tried and tried again and at last asserted beyond reasonable doubt. A hypothesis is set up in such a way that it can be confirmed or disproved; in this case that babies are able to distinguish, or simply *do* distinguish between men and women, and hence, even before they can speak, have an awareness of categories; and again, that babies can remember that they have already seen a photo and that this will be reflected in their brain reaction when they see it a second time.

And so on. Hundreds if not thousands of hours have gone into these research papers, each one of course sitting on top of scores of previous papers all of which are religiously and repeatedly referred to in the new papers, each one requiring a team of dedicated, highly educated people, using expensive and sophisticated equipment. Setting out for my interview, I suddenly feel daunted. Professor Pauen is an authority. I am just a guy who has had three children himself but can't really remember anything about their exact cognitive powers at nine months, or eighteen months, or two years. What chance do any of us ordinary folk have of figuring out where the scientists are really up to on matters of memory, awareness, consciousness? Wouldn't it be better simply to defer to their authority and have them tell us how it all works?

It's raining heavily. So it turns out my early morning intuition was right. The hotel reception very kindly provides me with an umbrella. I had checked the route I have to walk on Google, but now the Heidelberg streets don't seem to correspond with what I have remembered. I don't have a smartphone, but I did save the map on my computer which I have in my backpack, a strategy the philosopher Daniel Dennett loves to call 'outsourcing intelligence' and considers one of the great achievements of human evolutionary progress; you delegate information to a series of crutches – books, shopping lists, street signs – so you don't have to clutter your head remembering everything yourself. 'Supersizing the mind', another philosopher, Andy Clark, calls it.

Under a shop awning, then, I close the umbrella, pull out the computer and switch it on. I have to hold the thing in one hand and type the password with the other. It takes an age to fire up. I'm aware that my right shoe is leaking. I've had these shoes a long time. I'm the kind of person who hates to change things. My foot is damp. Then my phone rings. All kinds of juggling is required to pull the thing out of my jeans pocket. I could do with a real crutch here. It's the university in Milan. My employer. Am I going to answer? No. Am I going to be late for the interview? Very likely. There are philosophers and neuroscientists – many – who deny that subjective brain states exist. Anxiety, guilt, joy are just words we attach to our various behaviours. If the so-called subjective state can't be associated to an observable behaviour, or at least to some neuronal activity objectively trackable with an encephalogram or some sophisticated process of brain imaging, then we can rest assured it doesn't really exist. All the same, while I continue to behave the way any of us would in these circumstances, checking the map, identifying my mistake – I turned left a street too soon – scurrying under my umbrella, and cursing my leaky shoe, I'm intensely aware of experiencing an unpleasant cocktail of emotions, a corrosive mix of vulnerability, foolishness and guilt. Why guilt? Because it was presumptuous of me ever to imagine I could get my head round all this stuff. And anger. I'm wasting my time. I should never have got involved.

So, is this defeatist mood that's suddenly on me as I cross the courtyard of a rather noble old university building, a real *object* in the physical world? Does it or does it not, that is, have a physical existence? If it does, is it the result of an unhappy homunculus grabbing the controls in my command centre up top? Could it ever be located and measured at some precise position or positions inside my skull by some clever machine? Or is it, rather, as philosopher David Chalmers would have it, made up of some mysterious X factor, as yet beyond our ken? Whatever the case may be, this unhappy feeling, dense with negative premonition, is certainly what is most present and real to me – I feel bad, therefore I am – as I climb the stairs to the third floor of the Psychologisches

Institut der Universität Heidelberg; this even though I'm the only one who will ever know exactly how the feeling felt, and even though it will very likely have vanished by lunchtime. All this plus that damp foot. I hate having wet feet. In any event, I'm going to have to fight through this miserable mood if the interview is to go off successfully. There is, it seems, an I, or a something, who lies beyond defeatist moods and just gets on with the job.

First a broad old wooden staircase, then a broad wooden corridor. Polished wood. All very German. Pauen's door is on the left; I'm actually five minutes early. How can that be? Is my internal time observing a different regime from the official clock, the way my partner's and my internal temperatures never correspond? That little panic with the computer under the dripping porch seemed to take forever, but actually it was only a minute or two. If that's the case, was there nothing real in my impression that it took so long?

Since I get the feeling there are voices talking behind the door, I sit on a bench in the corridor and pull out my computer for a last glance at the list of questions I've prepared. A young woman comes to sit next to me. A couple emerge from a door opposite, cradling a yowling baby. 'Didn't enjoy his photo-recognition session,' I suggest to the woman next to me. 'Could be,' she smiles. It's fantastic how ready the Germans are to speak in English. Imagine addressing an English person in German in some similar situation and getting a prompt response.

Over the next few minutes this young woman tells me what hard work it is getting the babies to pay attention to the photos. But rewarding, because the results are so fascinating, the incredible responsiveness of their brains. When a couple appear from the stairs with a toddler she stands to greet them and leads them off to another room further down the corridor. At ten sharp, absolutely in line with what we agreed by email, I knock on the door and a voice calls me in. Sabina Pauen is alone in the room. Could she have been talking to herself earlier? Or on the phone? Or was I mistaken?

She is an attractive blonde woman in early middle age with a friendly smile. I sit down so we're looking at each other across a desk. Since neither of us is used to interviews like this, the atmosphere feels a little forced. More elusive subjective emotions; but how relaxed can you be launching into a discussion of consciousness at 10 a.m. with someone you've never met? Perhaps all serious reflection is wrested from a landslide of ordinary circumstance, is that the truth? And every book written against a tide of distractions? Even the neuroscientist denying the objective existence of sub-jective moods would not be immune from a bad temper over a damp sock. Though oddly my own discomforts have disappeared from the radar now as I concentrate on be-coming friendly with this woman, partly in the hope that she may tell me something interesting, partly because this is what I do when I meet people. I try to be friendly. It's an ambitious project, I suggest, that the Deutsch-Amerikanisches Institut has come up with. Does she ever have the impression that people are treating science like a religion?

'Certainly, they're always hoping I'll tell them something important,' Doctor Pauen laughs.

'They're looking for the reassurance of someone authoritative perhaps?'

With almost no preamble, we're right into our subject. Pauen tells me that various publishers, for example, are eager for her to write a book on her work, explaining how children establish categories for things, animals and people very early in life.

'They want me to spell out how the brain makes this first move towards concepts and categories.'

'Sounds like a good idea.'

'But I'm not really at all sure how it happens. So if I wrote the book I'd find myself nailed to a position that could easily be proved wrong. I'd look a fool.'

Wasn't this exactly my concern with the tea urn at breakfast?

'People want scientists to know things more certainly than they really know them,' I suggest. 'Is that it?'

'Right.'

I'm simultaneously disappointed and encouraged. Encouraged that Sabina Pauen is a pleasant and engaging person who has quickly dispelled my negative mood, disappointed that she is evidently not going to tell me something mind-blowing about babies forming concepts.

She asks me if I read the publications she sent and I tell her yes. All three.

'In fact I wanted to ask you about your use of the word "representations". You say in your paper that you're concerned to know how long it takes a child to *encode* a face and establish a *representation* of it.'

'That's right, longer than with adults. An adult will

recognise a face in milliseconds; a tiny child needs longer. Maybe a second, maybe a second and a half, at nine months.'

'But in either case, do you really mean there is a code, and a representation?'

This question provokes a silence. After a moment she says: 'These are the words we use to describe the fact that from this point on the child will react differently to the face, because he or she has seen it before.'

'But there isn't a representation in the child's head. A picture of the face in the head separate from the face seen on the screen.'

'No we don't exactly mean that.'

I'm puzzled. What *is* a representation, if not some kind of picture of something held separate from that something? What is a code, if not something used to produce a ciphered version of something else? Why use the words if this is not what they mean?

'So there isn't a code, and there isn't a picture, stored in the brain?'

This would be the right moment, perhaps, to bring up *Inside Out* where Riley's memories are stored in sealed, semi-transparent spheres, each about the size of a croquet ball, but with different colours indicating different emotions, and then stored in shelves in something resembling a vast, sparklingly clean, corporate archive. But I'm worried that if I mention the Pixar movie, Professor Pauen might think I was being disrespectful.

Meantime she is sighing and reflecting. There is no philosophical or metaphysical project behind her research, she tells me. She and her team are not trying to solve the ultimate problems of memory or of consciousness. What may or may not actually be happening in the head when these brainwaves occur, she cannot say. What does an N170 brainwave stand for, or an N160? Or a P1? Simply waves of negative or positive electrical potentials – what they call ERPs, event-related potentials – occurring a precise number of milliseconds – 170, 160 – after a certain incident, or 'event', like seeing a face. 'We can't say what that *means*, how that *works*, though obviously the neuroscientists have their theories.' She is a child psychologist, she insists. 'What we can say is that there is a pattern whereby when the nine-month-old baby sees a face a second time, the reaction is less marked than the first time. Or at least this is so when they see women, but not when they see men.'

'And you say that means they can distinguish categories. But presumably they don't know they're distinguishing categories.'

'In the various experiments we've done we've recorded different typical reactions to things and animals and people, and to men and women, suggesting an awareness of different categories. All this at earlier ages than people previously thought. So this is groundbreaking research.'

'Couldn't it just be that men and women and animals *are* different and hence evoke different responses, without the child having established categories?'

'But the thing is that they don't evoke those different responses at, say, six months. Then the baby doesn't distinguish. So in those three months something has changed and the child is now reacting to male and female faces with different intensities.'

'If you can get them to concentrate on the photos.'

'Right.'

To keep things cheerful I remark that I often find it pretty hard to distinguish men and women myself.

Professor Pauen laughs and agrees. She explains that before showing their photos to the children each image is examined by a number of psychology student volunteers who grade it for femininity or masculinity; any ambiguous or androgynous-looking faces are excluded.

'Do the students agree on their gradings?'

'For the most part.'

'So the babies are not actually distinguishing men and women, in absolute, but a cultural norm of femininity and masculinity as determined by a group of psychology students.'

'Correct, what matters is not whether a baby could tell a transvestite from a woman, but whether they have established the broad concepts, men, women, from their previous experience of those around them. That's why with this particular experiment we have stuck with Caucasian faces, that is faces from the same race and background as the babies' families. We just want to find out when babies become aware of these categories and, since they can't talk, we determine that through their neurological reaction to the faces. We also want to know whether they recognise a face, that is whether they're aware they've seen it before. And all the trials suggest, from the reduced amplitude of the brainwave response on second viewing, that they do recognise the face, they are in a sense already used to it.'

'But does recognition mean that the brain has *changed*? That seeing a face for an instant just once has altered something inside the brain, and hence it reacts differently the second time?'

'If you want to put it like that, yes. This is the brain's incredible plasticity. It has adapted so that it knows it has already seen the face.'

'But does the *baby* know? I mean, the baby's brain reacts differently, but is the baby aware that it's seen this face before and that it is reacting differently?'

'Nobody ever asked me that,' she tells me. She reflects. 'We assume it does. At some level.'

'OK, and would you say then that recognition is the same as memory? Is it "stored" somewhere?'

'Really we have no way of telling whether the baby might experience an unsolicited memory of the face.'

'And you don't know why you're getting the recognition effect with women's faces but not with men's.'

'At this age. I should stress that. Just a few months later children will have a recognition response to men's faces as well. But no. We can only speculate that perhaps they are more familiar with women's faces at this age, since the baby's world is mainly made up of women, hence they more quickly . . . '

' . . . encode it and form a representation.'

She laughs.

I ask her more generally about how the parents of the children see the experiment and see her. Won't it be a particular kind of parent who brings a child when they're receiving no money for doing so? Aren't they maybe hoping their child is ahead of the game? Or will get ahead of the game? Won't this effect the results of the tests? What if the children are not interested in looking at photos?

'Oh, they're often not interested at all!'

It seems there are any number of times when Pauen's team has to send a child home without finishing the experiment.

'The point is, though, that if we get twenty or thirty tests all going in the same direction, after all the maths has been done on the results, then we can publish a paper saying, at this age babies do distinguish this from that.'

'Though maybe not the ones who didn't get through the test.'

'Maybe not, no. But the important thing is to establish that some do, even at this early age.'

'What use is it all?'

The business of 'doing an interview' forces one to ask these questions. I'm aware as I ask it that I'm not really interested in the answer, only in how Pauen will field the question, even though neither of my briefs – the official 'is science substituting religion?' brief, and the unofficial 'what is the nature of consciousness?' brief – requires that I test Pauen in this way or think about how well spent her research funds are. I'm thus allowing a sort of cultural inertia to distract me: this is the kind of stuff a potentially indignant member of the public asks a scientist.

'It's a tricky issue,' Pauen acknowledges and she too immediately slips into automatic pilot, recognising and categorising my question just as her babies recognise and categorise male and female faces. Something in her brain was ready for this; she's been here before. As far as she's concerned, she says, her work is pure research, she has no goals beyond validating certain hypotheses and publishing papers about them. On the other hand, people are always asking her about practical applications, and when her team applies for research grants obviously it's tempting to make claims for some immediate practical use for the research, because that would make it so much easier to justify all the money it costs.

'I suppose,' she winds up tentatively 'that once we've established what is the normal path to development, at what age children make this or that distinction, at what age they learn to use this or that tool, or for example to distinguish the functional feature of a tool, its length, say, or shape, from some other attractive but non-functional feature, its colour for example, then we can know if a child is in the norm, more or less, and where each child is likely to be up to in their cognitive development at this or that age, and so, if we want, we could devise educational activities to take them in the right direction at the right moment. We can give them a bit of help.'

'Isn't there the danger that you'll just encourage anxious parents to worry and fuss? How is my baby doing, how can I help him get ahead?'

'There's that too,' she laughs, and at this point I realise I like her.

'By the way,' I tell her, 'the English in the papers is excellent. Is that difficult, doing all the research in German, then getting it into English?'

Oddly, this compliment born from the feeling that I rather like Sabina Pauen and would like to say something nice to please her, turns out to be my most fruitful question so far. The language problem is significant, she says, particularly when you're trying to publish something that flies in the face of received wisdom, so called. For example, she and her team are making claims about babies of nine months that were previously only made about older children, the fact, that is, that they can distinguish categories *prior to language use*, which means that one can have a concept of category without language. Something a lot of people contest.

'When you come up with results like this you get resistance from the referees reading the papers, people who maybe have made their names claiming something else, or have just been teaching something else all their lives. They don't want to know about research that tells them they were wrong. So they say your English is poor and that the article can't be published as it is. I think often they associate slightly incorrect language use with stupidity.'

This is definitely a reaction I experienced in my early years in Italy.

'There's an element of orthodoxy about science then, conservatism, dogma?'

'Oh, absolutely.'

'And in that sense it is indeed like religion. It doesn't want its beliefs challenged.'

'I suppose so. Even though the scientific method establishes that every fact has to be susceptible to the kind of experimenting that could prove it wrong. It has to be falsifiable, as they say, challengeable. Sometimes we've sent off papers we were convinced were absolutely watertight, with exciting results, and they got sent back with comments on the English. So it's encouraging, I suppose, but also depressing in a way, that you, a writer and translator, tell me the English is excellent.'

'It's good yes. I had no problem with it at all.'

There is no better way of making friends than inviting a person to air their grievances and agreeing they have been hard done by. 'I have a friend', I begin to tell her, 'who always has to fight like crazy to publish his stuff.'

But then before seeing how she reacts to Riccardo's seriously radical ideas, I decide it might be wise to invite her to lunch. She agrees at once.

No Images

'There are no images!'

This was the first time I noticed Riccardo Manzotti. He stood up in the audience and said those words most emphatically, in English, with a strong Italian accent.

It was September 2009, a conference on art and neuroscience at my university in Milan, one of those forums where people from different fields are invited to step back from the specificity of their particular disciplines and place their work in the context of more general human experience. The official title, 'Neuroesthetics: When Art and the Brain Collide', did not seem promising. Perhaps as a result, the atmosphere was unusually tense; the art historians and critics were anxious that they wouldn't understand a word of the scientific jargon, the scientists concerned that their work would seem underwhelming and wrong-headed.

Press-ganged into participation by my head of faculty, I gave a paper on 'The Matter of Words' reflecting on the difficulty of even imagining the manner in which language might adhere, as it were, to the brain and the body, and so become a fact in our physical experience. In general, I contended, language appeared to have a vocation for constructing a world apart, inviting us to abstract ourselves from the world as much as engage with it, to move in an environment of words and rhythms and syntax, separating our experience into two distinct categories, mental on the one hand and physical on the other.

In particular, having given words to everything we could see and touch – birds and beasts, rocks and trees, nuts and bolts – and then to everything we experienced as emotion – fear, hope, happiness, grief – we had also got into the habit of inventing words for things we'd never seen or even felt, but liked to believe existed anyway: angels, demons, fairies, God. 'Self' was one of these invented words, I suggested, an entity no one had ever really witnessed or grasped; so were 'identity', 'personality', 'character', 'soul' – the more words you have, the more believable the chimera becomes – thus creating the illusion that we, the modern individual, existed in some way apart from the physical world, not subject to the constant change to which the phenomena around us are subject, but rather projecting ourselves through time on a stream of words. Identity, in short, was a story we told ourselves; language and words were in strict alliance with an internalist, Cartesian view of reality: experience was all in our heads where we talked to ourselves, indeed talked ourselves into existence.

The paper drew some subdued applause, but then it turned out that none of the neuroscientists present had worked with language, with poems or novels, and hence had nothing to say on the matter. They were concerned with the visual, and in particular the brain's response to paintings and sculptures. The star of the show was Semir Zeki, professor of neuroesthetics (a word he himself invented) at University College London. His paper, full of PowerPoint slides showing complex graphs of neurological activity and murky images of the brain, focused on our reaction to ambiguity, a quality he considered to be at the heart of the aesthetic experience. A group of volunteers had been shown the Rubin vase – the black-and-white picture that looks simultaneously like a vase and like two faces in profile, their noses almost touching.

While the volunteers were looking, the usual sophisticated equipment mapped out the electrical responses in different areas of their brains. Sure enough, as each viewer tried to sort out exactly what he or she was looking at, a vase or two faces, so their brain activity flicked back and forth between two different areas in the fusiform gyrus (situated in the lower part of the brain), areas that have long been associated with object perception and face recognition. This demonstrated, Zeki said, that aesthetic responses had a precise neurological underpinning and that artists were, in their way, neurologists, exploring the possibilities of the brain in its production of visual experience and challenging us to respond in ever more subtle ways to our complex environment. Thus art contributed to, indeed *was* part of our evolutionary progress.

'Professor Zeki,' interrupted a certain Ron Chrisley, expert in artificial intelligence, just as the speaker was sitting down, 'if you tell me which circuits of a computer are active when its chess program moves knight to queen's bishop three, you really haven't told me very much about chess, have you?'

Even years later I recall how thrilled I was by this rebuttal. First its fearlessness, its being absolutely unimpressed by the findings of all those clever machines. But more than that the obvious point that there was an abyss between the subjective experience of the visual conundrum that the volunteers had been shown – the two faces, the vase – and the record of electrical activity in this or that part of the brain. How could one say the experience was *caused* by that activity, or somehow emerged from it, or even *was* that activity, just because the activity took place? After all, all kinds of other things were taking place, not least the ambiguous pictures themselves and the flickering eyes of the observer who looks at them.

But this excitement was as nothing compared to Riccardo Manzotti's sudden eruption into the open discussion that concluded the conference the following afternoon. There were about thirty or so people scattered across a room meant for a hundred, all weary at the end of eight hours of heterogeneous information from fields they were not familiar with. Zeki was again centre stage, talking about the contribution neuroscience could make to aesthetics, its ability to analyse in an *objective* way, he insisted, with *scientific equipment*, the process by which we generated the images we saw. At which point, Manzotti bounced to his feet and announced, 'But Professor Zeki, there are no images!'

He spoke for about five minutes. Everyone, he said, was focusing on what happened in the brain. Everyone was talking about input and output and information processing. Everyone imagined a subject quite separate from an object, as a result of which they had to suppose there were little pictures in the subject's brain representing the world outside the brain, little sounds in the brain, smells in the brain, colours in the brain, shapes in the brain, and so on. But none of this could be demonstrated. Scientists had looked for pictures in the brain and hadn't found them. They had looked for memories and hadn't found them. The brain was a greyish mass of billions of neurons and various chemical substances. It did not contain the world. If you closed your eyes the world disappeared. You could not cross the room you were in with your eyes closed. To have visual experience you needed the world. Aesthetic experience, like all other experience, was not locked up in the head; the experience of the *Mona Lisa was* the *Mona Lisa* as your perceptive faculties allowed it to exist when you stood before it. This was why people went to see the *Mona Lisa* rather than gloating over images of it stored in their heads, or indeed on their computers.

It was aggressive stuff. Manzotti had a rather wild look to him, an explosion of tangled hair, the most intense blue eyes. And he was passionate, ironic, derisive. He really couldn't believe how stupid we were all being, he said, buying into this dumb story of images in our heads. When he sat down I asked the girl next to me who he was. 'He builds robots,' she whispered. 'He's a genius.'

For some days after the conference I tried to test the truth of what Manzotti had said. I would sit down for a few minutes, close my eyes and try to establish whether or not there were images in my head. I had always supposed there must be. After all, isn't it entirely standard to say things like, I have this image of you when we first met. Or, I can still see so and so scoring that goal in the dying minutes. Or, her face is constantly on my mind. But do we, can we, is it? Could it be Manzotti was right?

Behind closed eyes I tried to conjure the faces of my mother, my brother, my children, people you're convinced you have an image of. And familiar places: I tried to visualise my office, the bench in our garden, the café where I go for a cappuccino. In each case I felt strongly that I knew what they looked like, but I didn't actually *see* them. As far as the objects and places were concerned, I could describe them to myself, in words. The desk to the right of the door. The bookshelf opposite. Perhaps, at a pinch, I could have produced a rough sketch of them on paper. I knew what colours the walls and the chairs were.

But I did not see them.

With faces on the other hand, it was rather like that moment at the arrivals gate at the airport when you are waiting for someone dear to you. I remember particularly at Heathrow once when my younger daughter, Lucia, was returning from Japan. She was only fifteen and I was terribly anxious to see her again after her month away. You stand at the railing as people stream through the automatic doors, searching for the face you love in the crowd of faces that mean nothing to you, willing for her to appear. You think of her intensely. You know perfectly well which face it is you are waiting for. *That* face. There is no chance you won't recognise her. It would be unthinkable. Something in your head is tensed for activation. You can feel it *physically*. When the face appears it will be as if a key turned a lock and a door swung open on full vision. Yet right up to the split second of the face's actual appearance, you do not really see it. The face is a tension, a potential. But you don't possess it. You can't produce it at will. That is the difference, I suppose, between presence and absence. Simply: when someone is absent you do *not* see them. That's the horror of absence. Or in some cases the blessing. There is no photo in the filing cabinets of the head. If there were, you wouldn't need ordinary photography. Quite likely there are no filing cabinets either. Manzotti was right. At least in the main.

What about dreams though?

I often *dream* my daughter's face. Then my eyes are closed and she is not physically there of course. All the same her face can be intensely present. In fact, the further away someone is in time and space, or at least this is my experience, the more likely you are to dream of them. I dream of my brother in America a lot. And when someone close to you dies, you are simply bound to dream of them. I dreamed of my father for years. Severe, or just distant in life, in my dreams Dad winks at me; he is complicit, though oddly he is almost always wearing his robes, his cassock and surplice and clergyman's dog collar. If you have to leave your wife, my father's face says in my dreams, then I am with you and I forgive you.

He would never have said that in life.

I found Manzotti's email address and invited him for a beer, the first of many. In the months and years that followed we went through our marriage crises together and he gave me an entirely different vision of what consciousness, or experience is, or might be. Plus a reading list a mile long, from the Presocratics through to Hume and Kant and William James, Ryle and Searle and Dewey and Nagel and Dennett and Ned Bloch and Varela and also lesser-known folks like Teed Rockwell and Alva Noë and Andy Clark and Mark Rowlands.

But where to begin?

With robots perhaps. Manzotti had begun by building robots. Or, to be precise, he had built a robotic version of the human visual system for use with anthropomorphic robots equipped with stereo eyes. He worked in several Italian universities, Genoa, Milan, Palermo, and later at the Korea Institute of Science and Technology. His main goal at that time had been to understand whether and how intelligent automatons might achieve something akin to human consciousness. It was here that he first realised, he says, that the standard model of conscious perceptual experience – input from without into a head that processes and computes – simply did not work. You could not build an intelligent robot on that basis. 'People say the robot stores images of the world through its video camera and compares them with the immediate environment,' he observed. 'But it doesn't, it stores digital data. It has no images in its circuits. No pictures. If there were pictures you would need someone to look at them.'

What were the implications?

Endless.

This is always the problem when I try to present Manzotti's views. It would be the same when I put them to Sabina Pauen over lunch in Heidelberg and likewise to the other two professors I met in the German town, the philosopher and psychologist Thomas Fuchs and the neuroscientist Hannah Monyer. People are in a hurry, Manzotti's claims are large and require a complete rethinking of who and what we are, and what the world might be. Even the apparently simple question, what is an object, needs serious revision. Not to mention the thorny question of time . . .

Then who is going to suppose that a theory they haven't heard of till now, something coming to them, what's more, through the conduit of a mere novelist, a man more famous for fiction than fact, might actually be substantial, or at least interesting? Isn't it easier to suppose that Parks's friend Manzotti is a charlatan, a non-starter, despite his many publications in serious scientific journals, his collaborations with prestigious universities, his PhD in robotics, and professorships in psychology and philosophy?

Or, alternatively, that Manzotti *is* serious, but that Parks hasn't really understood what he is talking about. Hasn't *begun* to understand. Often I fear that this is the case myself. Manzotti's ideas excite me, in many departments they convince me – unlike so many scientists he always makes a direct appeal to experience, he lives his ideas – but they leave me feeling vulnerable in discussion. They're so out of left field, though at the same time so commonsensical. Perhaps when I talk to people about his views I am really asking for my interlocutor to shoot these ideas down, to free me once and for all from this charismatic figure. I'm conflicted. No doubt if some neuroscientist were watching what was going on in my head they would see how Manzotti's ideas, like the Rubin vase, activate now one part of my brain, now another, the part that says something is true and must be urgently dealt with, and the part that says something is fantastical, preposterous and will bring only trouble and ridicule. Whenever I engage with Manzotti's thinking, my world feels unstable.

Which takes us back to the question of authority. Most people believe more or less the same things about the mind, the body, perception, and in most cases these are the things that various authorities would have us believe. There is hardly an ism, or a religion, or an intellectual elite, from Platonism through Christianity to empiricism and scientism, that doesn't warn us that special mental powers, special relationships with supernatural beings, or simply special and sophisticated machines are necessary in order to know what is really going on between ourselves and the world. Again and again it is drummed into us that only the clairvoyants, the geniuses, the priests, the scientists, the supercomputers really know. We, on the other hand, can't know. We are endlessly fallible. The result is that we regularly find ourselves signing up to explanations of reality that seem a million miles from our experience.

For example, when I suck a mint, it very much seems to

me that the experience is in my mouth at the meeting point of tongue, palate and mint. But contemporary science tells me that this is not the case, it is in my brain. 'The only reality we experience is brain reality,' writes Semir Zeki, 'the only truths we know are brain truths.' There are no such things as flavours except in my brain, says the great neuroscientist, pretty much repeating what Galileo said 400 years ago, even though, to date, the only things neuroscientists have found in my brain are a bewildering complexity of chemical transformations and electrical charges. Meanwhile the mint which smells of mint and tastes of mint stays, or dissolves, in my mouth and seems, while it lasts, terribly convincing as the location of my experience of mintiness. But who am I, or a mere mint, to defy so many authorities? Above all *scientific* authorities. And weren't they right in the end about the earth revolving and orbiting round the sun when to almost every ordinary person it seemed the ground was rock-solid still? To be honest, it *still* seems still to me today. Actually, *relative to my body* the earth is notoriously still and our senses are quite right when they perceive it as such. Only relative to the sun and other heavenly bodies does the earth move. Who, going about his ordinary daily business, would have thought?

So is there some who-would-ever-have-thought switch of perspective – as when I stop thinking about *me* and the earth and start thinking about the *sun* and the earth – that would shift the experience of the mint from my tongue to the so-called *nucleus of the solitary tract* (NST), an area of the brain located in the medulla in my head where an awful lot of electrical activity goes on when I'm tasting the mint, and again to the *ventroposteromedial* (VPM) *thalamus* where the electrical action then moves off and finally to two areas of the neocortex: *the insula* and *frontal operculum cortex*? For this is where the neuroscientists have tracked the taste experience down. Is there, that is, some quick change of point of view that would prompt me to agree, yes, looked at that way, of course the minty taste is located in the brain?

Manzotti says no, there is not. He feels that the fact that damage to these specific areas of the brain will damage your sense of taste does *not* mean that the experience of taste is located in the brain. It means that these areas of the brain form part of the equipment that makes the experience possible. Cutting off your tongue will also damage your sense of taste, while changing the food in your mouth is the surest way of tasting something different. The only place where there is actually mintiness, Manzotti insists, is in the mint in your mouth.

But it seems terribly unsophisticated, the wisdom of the fool in fact, to say these things to accomplished psychologists like Sabina Pauen, or acclaimed philosophers like Thomas Fuchs, who I am going to see in the afternoon after my lunch with Sabina Pauen, or prize-winning neuroscientists like Hannah Monyer who I am going to see tomorrow.*

How to begin, then, with these authoritative people, having discussed their own work, to open up the discussion to

---

* I have since been comforted to learn that the philosopher Thomas Nagel observed that supposing one had access to the brain of someone eating chocolate, supposing one could lick the appropriate part of the brain, it would not taste like chocolate.

what most interests me? One is so afraid of seeming stupid, even more so than with the maid at the breakfast table over the question of the tea urn. And writing this down now I am afraid of seeming stupid again. Parks has gone out on a limb, and he doesn't even have the cover that it's fiction.

To make matters worse, Manzotti has changed his position since I met him in 2009. Changed it radically. At the time he was talking about experience being a process shared between subject and object, a sort of *pas de deux*, where the embrace is guaranteed by the uninterrupted continuous movement of photons or sound waves or odorants travelling between object and subject. Or, in the case of touch and taste, by contiguity.

His preferred example in those days was the rainbow. When the sun is low enough and its rays pass through a cloud in which a large number of raindrops are suspended, an observer finding him or herself in the right place at the right time will perceive the refracted light as a rainbow. If there is no observer at the right point, there is no rainbow, just drops of water and light. For example, a scientist cannot describe, let alone measure, a rainbow without knowing the position of the observer, since it is the observer who determines exactly what rainbow is selected from the huge mass of raindrops through which the sunlight is passing. An observer in a different place sees a different rainbow or none at all. This is not subjectivity. It is a question of being in a different position with regard to the raindrops and the sun.

Manzotti argued that all visual experience was of this nature, carved out by the physical processes set in motion when the nervous system encountered the world, a collaboration of thousands of elements and brain activities spread across the entire distance between what we wrongly call subject and object. The experience was the whole caboodle. The mind was spread between the body of the perceiver and the perceived external object. In fact it was around this time that he started calling the theory 'the Spread Mind'. Shut out any part of the process and the experience immediately lapsed. The sun goes in and the rainbow is gone, even if the raindrops are still up there suspended under their cloud. Close your eyes and again the rainbow is gone. Move position and it disappears. Lose your eyesight and there'll be no more rainbows. For you.

'We'll come to dreams later,' Riccardo would say when I raised that objection. 'We'll come to hallucinations later, don't worry. Let's deal with the day-to-day stuff first.'

But just as I had begun to accept this approach – experience as process – and in the meantime Riccardo's wife had left him, because of his many relationships with other women you might say, though he would say that in fact he and his wife had long been in an open relationship, something that she herself had encouraged, since quite likely she was already planning to leave him, so that as with the observer and the sunshine through the raindrops it was hard to say which exactly was the cause and which the effect, who the victim and who the perpetrator – just, I was saying, as I had begun to accept this approach – and Manzotti, reassuringly, was not its only proponent; famous names like Gibson, and Bateson, O'Regan and Noë had also set off in this direction – just as I had got used to feeling, as I walked down the street, that my experience of sky and buildings and spring smells and traffic noise, hard pavements and soft verges, was not simply in my head but also outside of it, spread across space and time, a wonderfully busy collaboration of processes at once all around me and within, he changed his mind. This version was wrong, he told me, and I must no longer read any of his publications pre-2015. Experience is *not* a process. He had been wasting his time for ten years and more.

'Why?'

'Because the process, which of course, don't worry, is definitely taking place, as described, *doesn't correspond to the properties of the experience*. So the experience can't be the process.'

Riccardo's style is this: we've eaten a pizza, we've talked about our marriage woes, or perhaps some likely woman we've met, someone who might promise a new direction; then quite abruptly he turns over the sheet of paper that served as a placemat and rapidly starts sketching diagrams on the other side. He's good at diagrams, good at sketching rainbows and people watching rainbows and apples and later, as we shall see, lakes and dams, and brains and neurons and the insides of your eyes and ears and all the battery of paraphernalia (he posts comic-strip versions of his theory online) that he has built up over the years to convince people of his ideas, including, it has to be said, some folk in very high places, MIT, Harvard, Cambridge, who are nevertheless all rather nervous about coming out and saying that Manzotti might be right, might have something going for him. Because the truth is, at the end of the day, that it is not the person who is right who is really right, if you see what I mean, it is the person who *convinces* everybody he is right. At least to all intents and purposes. Meaning reputations and research funds. It is no good being right *on your own*. No one will give you money till you convince the others.

'Forget rainbows,' he announces, 'and take an apple, a simple apple.'

He drew one. With a generous flourish of leaves on top.

In parenthesis, I should say that this is the only pub in Milan that serves London Pride, shipped in on a regular basis by truck under the Channel, across the Alps and down into the North Italian Plain. As a result, I have a special feeling of home here – the place is called La Belle Alliance – something I feel in my mouth and gut, as if part of my London adolescence had reappeared in Italy forty and more years on. It's a very physical thing and as I step towards the bar and order a pint of Pride (because they use pint mugs here, not metric measures), I won't say I have the taste already in my mouth, no, but I do have that wonderful *anticipation of recognition*, as with Lucia at the airport; the taste is *about to happen* and I am about to be taken back to London in the 1970s. The one problem being that often they put the music on too loud, so that presently we are being bombarded by sound waves from Creedence Clearwater Revival – more memories – we are rolling down the river, Proud Mary, as Riccardo explains his new and he feels sure now, final position.

'So, you're looking at an apple. OK? Let's call your experience E.'

Like everyone who has studied philosophy, Riccardo always feels the need to denote things with letters – P, Q, X,

Y – and, like anyone who has no formal background in logic or philosophy, I experience this as confusing, unnecessary, and a deliberate attempt to set up a teacher–pupil relationship where I'm the pupil of course. My conversations with Riccardo are also a struggle with Riccardo. I'm not going to cast off blind belief in one set of authorities, just to be in thrall to another.

'What are the properties of E?' he demands. And without waiting for a reply, announces: 'Red, round, and applish. OK?'

I'd like to find something to object to, but can't. It's indisputable. That's our experience of the apple.

'A bit of green,' I suggest. 'A bit of shine.'

'That's what I meant by applish!' He frowns. 'Now we have agreed, haven't we, that experience must be physical and hence that it must exist *somewhere*? Right? We're not going to accept the idea of some mysterious spirit or substance beyond our ken, as per David Chalmers. Nor any Cartesian ghost in the machine. No invisible stuff nobody can see. Nor any magical supervenience, courtesy of Dennett and Bloch and Clark and company, consciousness as something that just happens, like a genie arising from a bottle, because there is a lot of neural activity going on. Agreed?'

'Agreed,' (reluctantly). The truth is I'm genuinely despondent about having to jettison the notion of process. It seemed to make sense. I had got used to it. It was out on a limb, but it seemed a scientific and easily explicable limb to be out on. Experience as process sounded credible and respectable. Process is not a word you need feel ashamed of. So much of what we believe has to do with habit and convenience.

'Good!'

Riccardo, I should say, is a big man who runs and works out regularly. He has a strong, meaty presence, so there is almost a physical as well as rhetorical coercion when he gets going. But this is encouraging. He never forgets that we are here in the flesh and that this stuff he is talking about is the stuff of us being here, rolling down the river as Creedence are still crooning.

'So, given these applish properties of my apple experience, what candidates do we have for the location of that experience? I can think of three.'

And he writes down the list on the back of the pizza placemat.

N = neural activity.
O = the apple, the object.
P = the process which is a constant chain of activity connecting and involving **N** and **O**.

'So what's our best choice?'

He looks up and grins.

'I'd be happy to stay with the process, Riccardo. That's where my heart was.'

He shakes his head.

'Let's take 'em one by one. Candidate N. Neural activity is grey, bloody and gooey, right? It is not round and applish and red. You can look in your head as long as you like and you won't find apples or anything like apples or images of apples or smells of apples, or supervenient apples, or magical apples.'

'Bad choice, then.'

But this is something we agreed on ages ago.

'On to Candidate O, then, the object, in this case an apple. Well, apples are very like apples and very like our experience of apples, are they not? They are extremely applish.'

It's hard to disagree with this, but I still haven't seen where he is going. I'm briefly reminded of that scene in *Indiana Jones and the Last Crusade* where Indiana has to choose which of various cups is the Holy Grail. We mustn't make a mistake here.

'And finally Candidate P, the process. What do you think? This candidate is immensely complicated, even more complicated than the neural activity, because it includes the neural activity as well as a whole host of other things. We've got the surface of the apple reflecting light rays, we've got physical quanta bombarding the retina, hyperpolarised at the receptors, depolarised beyond, we've got nerve impulses fizzing up into various areas of the brain courtesy of a million synapses, then the neurons all doing their busy stuff. Is that like your experience of the apple?'

I don't know what to say.

Riccardo sighs. 'If we record what's going on in your brain while you watch the apple we'll find all kinds of fluctuations and oscillations. Is that your experience of the apple? Does it fluctuate and oscillate?'

'No.' I feel I'm on safe ground here.

'When you see the apple, it stays right where it is.'

'It does, yes.'

'Steady.'

'Steady.'

'It's not going to roll around unless we give it a push.'

'No.'

'At the same time, your pupils, which are part of the perceptual process, are constantly flickering back and forth. But you don't experience that movement.'

'No, I don't.'

'When you're seeing, you don't experience your eyes seeing.'

'I don't think so.'

'You see the apple.'

'Right.' I'm looking at his sketch on the placemat.

'You don't experience your brain.'

'I don't seem to.'

'In fact the brain, where all this experience is supposedly going on, is one of the parts of the body you experience least. Am I right? You don't feel your brain the way you feel your arm.'

'Only when I have a headache.'

'That's another question for another time. For the moment, you experience the apple, not your brain.'

'Yes.'

'So which of our three candidates – N, O or P – best fits the properties of your experience of the apple?'

Silence. Or rather Creedence. They were playing 'Through the Grapevine.' And isn't it fascinating how experience is never entirely one thing, never entirely focused, say me reading Kant, but always multiple, me reading Kant with the sound of Eleonora drying her hair, me reading Kant with a smell of apple pie fresh from the oven, me reading Kant with a backache from bad posture. Or all of the above. And now Creedence wailing that someone wasn't going to be someone else's for very much longer.

'The fact is,' Riccardo goes back to his paper and begins sketching rapidly, 'if we choose N or P' – and beside the letter N he puts together a tangle of neurons, and beside P a face, a fizz of photons, an apple – 'then we're going to need some pretty elaborate justifications as to why something that is very far from being red and round and applish is experienced as being red and round and very applish. Right? We're going to be stuck with the old conundrum, the brain looks like this, the process seems to be that, but the experience is the apple.'

'And so?'

Riccardo laughs. The music changes to something I don't recognise. 'Didn't Sherlock Holmes say, when you've dismissed all other candidates, the most unlikely one has to be right?'

'Something like that.'

'So, your experience of the apple is identical with the apple. It has to be! Only the external object is identical with one's experience of the external object.'

This is like being invited to step off a conceptual cliff, trusting that there'll be something beneath your feet when you do so. Again I'm reminded of Indiana Jones. His leap of faith.

'You're saying *I'm* the apple?'

'I'm saying your experience is the apple.'

'But the apple is there and I'm here.'

'Your body is here. Your experience is there. Your body is in one place, your experience is where the apple is because the experience and apple are one. Experience is identical with it, thanks to the causal process linking the object to your body. You are apple. Identity.'

'But the apple weighs something and I don't feel the weight. The apple is juicy and I'm not tasting the juice. The peel has a certain texture and I'm not feeling it.'

Riccardo speaks in the tone of one addressing a four-year-old. 'I'm not saying that your experience is identical to an *ideal* apple, or a Galileo apple, or an apple as God sees it, or a Kantian, noumenal apple, or an apple in an X-ray machine, or a slice of apple under an atomic microscope, or a chunk of apple in your mouth. And I'm not saying that your experience of the apple wipes out Creedence and the grapevine and your haemorrhoids and worries whether women are interested in you, and a whole host of other experiences and identities. I'm saying your experience of the apple is identical with the apple that your perceptual equipment carves out when it looks at the apple, with your eyesight, not mine, and your brain and your past experience of apples of various kinds and these light conditions and this particular apple. Only the apple, that apple, now, satisfies the properties of your experience, now, of the apple. The experience is the apple you experience. Take away the apple, no experience.'

He grabs the placemat, wobbles it up, laughs and drains his beer.

The apple is gone. But a powerful turbulence remains.

'I'm going to dream it,' I tell him. 'Fresh off the Tree of Knowledge. What about dreams?'

'Dreams later! Hallucinations later! Don't worry!'

Midlife

'What about dreams?' Sabina Pauen asks.

It has taken quite a while to reach this point. No sooner were we outside the university than the psychology professor seemed to lose all interest in the brain and memory. Now she was any happy mother eager to talk about her family, or generally get on with her busy day. And then there was the rain, of course, which immediately reactivated my damp foot. My foot must have been damp throughout the interview with Professor Pauen, but the discomfort had disappeared while we talked, no doubt in response to the effort of concentration one makes in situations like this. Certainly any model of the mind and experience has to consider the question of focusing, of how it happens that some experiences are, as it were, uppermost and others thrust into the background. In his *Quest for Consciousness* Christof Koch asks us to envisage a competition in the brain between competing coalitions of millions of neurons, each putting forward different potential experiences – a puddle to avoid, a voice complaining, a letter that needs answering, something you want to buy – each circuit firing off electrical charges at the synapses – the connections between the neurons – in an attempt to produce the critical level of energy that will place it at the centre of attention. It's a mechanistic, who-shouts-loudest explanation, but easy to imagine and very much in line with Pixar's *Inside Out*, the charming homunculi simply substituted with unruly crowds of neurons. David Eagleman, in his book *The Brain, The Story of You*, takes the same position. Every choice is a battle between 'warring networks of neurons'. Neuroscientists have recently been comforted by research that suggests neural patterns change when an experience moves from the periphery to the centre of consciousness.

Whatever the case, no sooner are we out in the street with the rain teeming down than my unhappy foot makes its presence felt. It's positively squelching. I'm going to have to buy new shoes. In a foreign town. Pauen, meanwhile, is concerned about how much time she has before an appointment with her daughter, who is shortly to move house and is calling on her parents to help decorate her new apartment. Which restaurant to choose, Pauen wonders? I tell her I'm vegetarian. That can be quite a complication in carnivorous Germany. We set off along the crowded Hauptstrasse, looking for somewhere serious. But raindrops are splashing on the cobbles. The Hauptstrasse is awash. The temperature is falling. Autumn has arrived. Our umbrellas are no match for the downpour. We take refuge in the nearest café not a hundred yards from the Psychologisches Institut and have the good luck to be given the only available table in a tight corner.

At this point just five minutes have passed since we were

sitting conversing about encephalograms and toddler attention curves, and yet in terms of experience it feels a great
deal longer than that, as if defending ourselves against the
rain and weighing the pros and cons of going further and
faring worse had actually taken an awfully long time. The
impressions have been intense; first the unpleasant trickles
off the sides of the umbrellas, then conversation and indeci-
sion in the bustle of the noisy street, finally the brief flight
to the café and the sudden change of atmosphere as we
entered its steamy warmth and tried to cram our dripping
umbrellas into the umbrella stand. If there is one thing that
is not a reliable measure of chronometric time, it is human
experience. Or to put it another way: if there is one thing
that is not a reliable measure of human experience, it is
chronometric time.

But the chronometer is vital when arranging meetings.
Professor Pauen was supposed to be seeing her daughter at
1.30 and it's already 1.10. She starts to make phone calls
while I study the menu and happen on the *Schafskäse*, a very
rare case of my knowing both sides of a German compound
word, this thanks to O levels in my adolescence and three
months working on a farm in German-speaking Switzerland
shortly thereafter (where there were no sheep, however). For
forty years and more traces laid down by exposure to those
words – printed in my school textbook, as I recall, in
Blackletter Gothic – have been silently waiting for this hand-
written menu in a Heidelberg café. If you had asked me this
morning what was the German word for sheep I couldn't
have said. But when it turns up on the menu, I recognise it.
On the other hand, I could have told you what the word for

cheese was; *Käse* is there when I want it. So there must be some difference in the quality of those traces. If 'trace' is the right word.

These reflections might be a good way in to the conversation on consciousness, and above all Riccardo's ideas about consciousness, that I'm now eager to have with Sabina Pauen. To get her opinion. Or at least reaction. Except that for the moment there is so much else to be conscious of that I miss my cue. Not least our extraordinary waitress. Wearing the full German traditional peasant costume, this matronly lady looks like an extravagant advertisement for wholesome values and old-fashioned hospitality. Over breasts frothing with milky frills, her blonde hair is half hidden in a bright red headscarf. When I say '*Schafskäse Salat, bitte*,' Professor Pauen compliments me on my accent.

'Don't ask me to write up our interview in German!' I tell her.

Ordering chicken, she asks me how long I've been a vegetarian and why. It's one of life's curiosities that those who don't eat meat have to explain why not, while those who do consider themselves the norm, hence no explanation is required. The norm always trumps unusual behaviour. And unusual ideas. Somebody who believed that the earth revolved round the sun in the sixteenth century was almost obliged to become a zealot, or a closet eccentric. Likewise anyone who believed in votes for women in the eighteenth century. I explain to Professor Pauen how I once served in the kitchen at a meditation retreat cooking strictly vegetarian food for 150 people. One lunchtime, washing up the pans, it occurred to me how many chickens and lambs would have had to be butchered and roasted if the meditators were meat eaters, how much blood and offal would be involved, how greasy the oven and the dirty dishes would be. Without really making a decision, I tipped over into vegetarianism.

Needless to say this is an anecdotal simplification of a more extended and complicated process of behavioural change. But it will have to do, or we won't talk about anything else.

Pauen is interested in meditation, she says, and ideas like karma. Her work with children has led her to believe they already have a self, or personality, at birth. She asks if I could recommend a meditation retreat and what I think about research that has shown that practised meditators show high levels of gamma waves during meditation.

This is my chance.

'Could you tell me,' I ask, 'exactly what gamma waves are?'

From this point on I will be deliberately piloting the conversation, I've seen a path to where I want to get to. Like my mother, I'm a tireless manipulator. It's a bad habit I picked up early in life, along with the rhetorical strategies of my father's sermons. Fortunately, this effort to draw my psychologist into a reflection on the Spread Mind theory finally pushes my damp foot into the background again.

Pauen explains that gamma waves are the fastest oscillations in the brain with the lowest amplitude; they are associated with high levels of concentration and feelings of well-being.

'And you would like to meditate because neuroscientists have found these gamma waves in Tibetan monks?'

'We could all do with relaxing,' she laughs. 'Except it's hard to see how I could get away for ten days.'

I suggest that there is an interesting question of authority here, something that relates to her own work and to the project I am involved in on behalf of the Deutsch-Amerikanisches Institut.

'A few years ago', I elaborate, 'I talked to a scientist called Ron Chrisley. After a conference. He had been measuring brainwaves in meditators, and said the results were interesting. There was definitely an alteration in the gamma waves. I asked him if he'd ever meditated himself, and he said, no. So I suggested that maybe he would know more about meditation if he meditated, than if he measured brainwaves while other people meditated. He would feel the benefits, or otherwise, himself. At which point he might be less interested in the gamma waves. He would simply know what meditation was, at least as he had experienced it, even though he wouldn't be able to publish a paper about it.

Pauen agrees. Science sets itself up as an authority superior to our own experience. It tells us what is good or bad for us. On the other hand, this is what people seem to want. She is always being asked, she says, to give people certainties about things no one understands. For example, gamma waves. Nobody really knows what they're about, or why they occur when there is a state of well-being.

'My brother,' I tell her, 'has various back problems. I proposed he try shiatsu, which I have found useful for all kinds of aches and pains. But he refuses. He says he did some research online and "the science isn't there". So even if shiatsu made him feel better, it might only be a placebo. As if, without science, we would distinguish between a placebo effect and a "genuine" improvement. We would just enjoy feeling better, something that surely has its own reality.'

Again she takes the point. There are a large number of pharmaceuticals, she says, that work without our knowing how they work. Since she seems so amenable, exactly as our plates arrive and a bottle of sparkling water, I hit her with Riccardo's theory.

'He calls it the Spread Mind,' I conclude after perhaps eight or nine minutes of monologue, 'or the mind–object theory. That is, rather than considering the mind more or less identical with the brain, or arising from it, as in standard neuroscience, or again as identical with the whole body, as in one or two other theories, or the body and the things it interacts with, as in enactivisim, he considers the mind, or conscious experience, as identical with the object. The brain and body, in contact with the world, allow experience, or mind, or consciousness, to happen, but the experience is not located in the body or the brain. It's outside. The object is the experience. This *Schafskäse*, for example.' Which, with all the talking I'm doing, I haven't even begun to eat. 'My experience isn't *of* this white lump of cheese. It is the white lump.'

Pauen chews her chicken. I'm wondering if I've lost all credit. Finally, she says. 'Well, if this were true, there would be one huge advantage. We wouldn't have to talk about representations in the brain.'

'Right, when you show the baby a face, something happens in the brain, but what happens is not an image of the face. And when the baby sees the face a second time, it doesn't compare it with a gallery of stored faces. The face fits an impression it caused to be made when it was first seen, like a key fitting a lock. The brain was predisposed to the reappearance of the face.'

I begin to feel a little pompous as I say all this, a little insecure, as if not sure why I want to insist. I turn back to my food. The Germans certainly know how to make a good salad. Not to mention that dark bread.

'So, the idea is that there is no experience without an external object experienced? Is that right? You need something outside to fasten on to.'

'Yep. Or to put it the other way round, if you're experiencing something, be sure it's there.'

'What about illusions, then? What about dreams?'

'Good question.' I give her a big smile and wipe my mouth. 'Are you ready for it? You'd better hold on to your seat.'

'Oh yes? Why?'

She smiles.

'Because this approach requires a major revision of ideas of space, time and a whole lot of—'

Somebody calls across the crowded room. Professor Pauen looks up. It's her daughter. Beaming, the mother stands and rearranges the chairs in our tight corner so that the young woman can sit beside us. We order strudels and coffee. They chatter in German. And I'm spared, or denied, my explanation of dreams in the Spread Mind theory. Walking back to the hotel on the wet pavements with my soaking foot, I wonder what Sabina Pauen thought about Riccardo's ideas. Certainly she was polite. Certainly she saw the

80

positive side, that it would free neuroscientists from searching for representations in the head. A hopeless quest. On the other hand, she was clearly underwhelmed. As if this were *just an idea.*

Those words force me to stop and take a deep breath. Just an idea that my experience of the Hauptstrasse right now might be outside my head rather than in, identical with Starbucks and Runners Point and Buchhandlung Schmitt & Hahn, not to mention the passers-by hurrying under brightly coloured umbrellas, the cold air on my face, the damp under my foot. None of this fabricated in the brain, but a collaboration between body, brain and everything outside. The world not in two places – colourless out there, and colourful in the thalamus – but one single kaleidoscopic happening.

All this *just an idea*! Nowhere near as important as showing babies photographs and trying to figure out whether they can determine categories at age nine months or twelve. Nowhere near as important as your daughter's new apartment. So whether the Spread Mind theory is true or false, or declared to be true or false, the world will go on much the same anyhow. Including the world of behavioural and cognitive research. Can that really be the case? The way everything went on just the same after it was officially settled that the earth revolved round the sun. People went on talking about the sun rising and setting, as if nothing had happened. And in the end they were right to do so. Because to all intents and purposes, from our point of view, the sun does rise and set, it goes up and down, it's high in the sky or low, and one can't always be thinking of the world in terms of astronomical absolutes. Common sense has no problem saying the sun also rises.

But is the world, from our point of view, inside the head, as they've been telling us for so long? Does it feel like that? Is the mind/brain identity theory that 99% of the world subscribes to – all your experience is locked in your skull – as self-evident, from our point of view, as the impression we have of the sun moving in the sky? Or is it rather a counter-intuitive idea imposed on us by a certain way of thinking, a centuries-old orthodoxy that has certain advantages, but also enormous disadvantages. To put it as bluntly as possible, would a person who has never studied philosophy or neuroscience *naturally* suppose that experience was shut away in his or her head and only in approximate connection with some kind of world outside? The way they would *naturally* suppose the sun rises?

Think.

Imagine that we have no information to go on but our own most basic day-to-day experience.

Well, when we see the world we know that the eyes must be the connecting point between visual experience and body, because if we close them the world disappears. But do we feel that the world is somehow projected into our head through our eyes, the way a camera lets a view in through its lens, and is then separately processed there into representations and images so that we have one thing in the head and another outside? I don't think we do. Processing and representations are science talk. To me it feels like the Hauptstrasse is out there, not in my head behind my eyes. The Byzantines, following Empedocles, believed vision was consequent on rays projected *out from the eyes* into the world, which was why they drew and painted so differently. Even the eliminativist, Daniel Dennett, a man who doubts the very existence of consciousness – eliminates it, in short – pointed out in an early paper that without the 'aid' of science *we have no way of knowing where our experience is.* We don't feel the Deutsches Verpackungs-Museum, or the onion dome at the end of Heidelberg's Hauptstrasse *inside our heads.* Rather we have the distinct impression, illusion, Dennett would say, that they are out there, the museum to my left and the dome straight ahead, rising above other buildings some 300 yards away.

Who would ever want to visit a packaging museum? In the head or out?

Likewise, when we hear noise we know our ears are crucial because if we stop them with our fingers, or in my case, while I'm working, with bright red foam earplugs, the sound changes. All the same we don't think the music blaring out from Parfümerie Douglas is being, as it were, *replayed* in our brains. On the contrary we feel that this noise is to our left, or that the sound of that rapidly accelerating car is coming from behind us. So we'd better move fast. For all its billions of neurons and trillions upon trillions of internal connections fizzing with electricity and chemical change, the brain seems mute and transparent. Riccardo is right about that: if there is one thing you don't feel when you're perceiving the world, it's your brain. Everything seems out there.

So why sign up to the standard theory?

Just because the various authorities have told us this is how it is?

Or because it is attractive to think of our mind, our experiences, as entirely ours, as separate from the world experienced?

And why does all this seem to me to be a matter of utmost urgency when to Professor Pauen, who actually gets her living from studying the brain, it's not so compelling? 'Hold on to your seat,' I told the cognitive psychologist, 'this approach requires a major revision of our ideas of space and time.' But she was immediately and irretrievably distracted by the arrival of her daughter who is moving house and needed to talk about a list of things to buy at IKEA. That far my rusty German got me.

Interesting hearing how differently IKEA is pronounced in different languages.

Why is it, then, I try to decide, walking back to the hotel where Eleonora is waiting for me – we exchanged texts after I arranged to have lunch with Pauen – why is it that all this seems so important to me? Why do I feel it ought to be important to others, or at least *might* be important to *some* others? Important enough for me to want to write about it when I would perhaps be better off writing novels.

It has to do, perhaps, with the changes in my life over the last ten years. My falling into and struggling out of a chronic illness. My decision to leave my marriage. My moving to live in a small flat in the big city of Milan after decades in a handsome house just outside the small town of Verona. And, more recently, my relationship with Eleonora. In each of these cases there was an official narrative that I eventually chose to resist, a conventional wisdom that contrasted with actual experience. You should be doing this, this and this, the various authorities told me, when all my senses told me I should be doing that, that and the other. In each case, whenever there was unease and anxiety, perhaps pain, perhaps unhappiness, certainly confusion, the only salvation was to try to see things *as they really were*, to give credence to one's experience and act on it, rather than persevering with the official version and customary caveats in the teeth of the facts.

But how difficult it is to see things as they are!

You go to a doctor about an embarrassing and painful complaint. The doctor says your problem is such and such. He has seen it a million times. He will have to operate. Nevertheless, *as a precaution*, he runs a battery of expensive tests. You are exposed to X-rays, while peeing. Ultrasound waves are radiated inside your anus. A clever tube with a video camera is pushed deep into your penis. All bodily fluids are analysed and examined. So now we have a battery of objective data. We have numbers, percentages, long lists of proteins and bacteria, graphs and bar charts and shadowy images on transparent plastic. Which all confirm his first impression, the doctor says. He will have to operate.

One's instinct, at this point in the story, is to accept the medical man's authority, the authority of hundreds of years of accumulated scientific knowledge, the facts as determined by sophisticated electronic equipment, and to feel relief even that we now know, regardless of my actual experience of the symptoms, what is wrong. We have a diagnosis. I remember my father's relief when he saw the X-rays of his lungs, X-rays that condemned him to death. At least he had an explanation now. At least he knew. He no longer needed an excuse for feeling so bad. Science had legitimised his breathlessness and his fatigue, his unwillingness to climb into the pulpit. Knowing he was seriously ill, he could stop trying to behave as if he wasn't. I can stop trying to be cheerful, he told me rather cheerfully.

The science was dead right with my father. The computer printouts and shadowy images fitted perfectly with his experience. All too soon I, my mother and my sister would be riding in a big car behind his hearse. But my situation wasn't analogous. I had a long list of symptoms that went far beyond the condition the doctor was eager to operate on, symptoms that the objective data didn't account for or coincide with. I took this data to another doctor, a more expensive doctor, and he agreed that my situation was by no means as clear-cut as the first doctor suggested. It was open to interpretation. I tried to explain to this second doctor *exactly* what my experience was. Unlike the first doctor, he really listened. For about twenty minutes. He was being paid a lot more. And he said he really had no idea what my problem was. I was on my own.

Similarly with marriage. For years you compare your married life with received wisdom on the subject of marriages. For years you try to understand whether it is normal that a marriage be as unhappy as yours is. You read novels, newspaper articles, sociological studies. You take advice. This or that strategy might improve the situation, you are told. Of course the problem is that the proposed strategy has to be something you want to try. That you both want to try. And somehow it always seems that when one does, the other doesn't. Some perverse mechanism clicks in; you are never in sync. You consult friends and even experts who now give shape to your unease with the concept of midlife crisis. Like the symptoms of a disease confirmed by a blood test or X-ray, this diagnosis is supposed to reassure you. It is *only* a midlife crisis. It is not real, not important. Certainly not permanent. You're not going to die. Just as colours are generated in the brain, so unhappiness is locked inside your midlife head. It's an illusion. The smart thing would be to accept the version of long-term reality that the experts and received wisdom offer you; which is to say that a midlife crisis passes, that decisions taken in response to a midlife crisis are decisions taken on the basis of illusion. Your dream of a different happier life is not unlike a mirage, a man hallucinating oases in the desert.

'So and so left his wife for someone absolutely similar but twenty years younger and now finds himself in exactly the same situation as before except with a whole load of extra expenses and unpleasant emotional fallout. For the kids particularly.'

This is the kind of story people tell you. Your wife backs it up. Your wife supports the conventional view. Your suffering – yours *and* hers, for she is suffering just as much as you are – is not real suffering but a midlife crisis. Not something you solve by taking action, but by waiting for the midlife cloud to pass, waiting for the projected marriage narrative to resume its textbook trajectory. So while you want to seek counselling, she wants just to soldier on. With the official version. The midlife crisis will pass. Like the common cold. Like a bad smell. The narrative now foresees old age and grandparenthood. In the long term, received wisdom will prove right. That's why it has become received wisdom. You will be happy in your twilight years. Hopefully some time before the shadowy X-ray that foresees the end.

For years you accept the logic of this. For years you seek to attach as little value as possible to your experience. Your unhappy experience. It is in your head. Outside your head, *objectively*, you are well off. Happy even. You have a nice house. Your work is going well, or well enough. And of course there are things like children and financial commitments and reputation.

Reputation! What our mothers think of us! And maids in three-star hotels.

Yet whenever in a careless moment, you find yourself noticing how life really feels, what your symptoms actually are, then you are aware that they began long before alleged midlife. And actually you find it difficult to see yourself as someone who acts rashly on whims and fantasies. If anything you seem to be the kind of creature who reflects and reasons a little too much, who works a little too hard to persuade himself that all is well and all manner of things shall be well, when it really does not feel well at all, on the contrary, you have the distinct feeling that if you don't do something drastic, this unhappiness will go on forever. Somehow, your circumstances can't be bent to fit the standard diagnosis. Once again, you are on your own.

And so on. The same for the move from handsome house to small flat, the same for the decision to get serious with a woman half your age. Less than half. 'She's nice, but rather young for you,' your mother said when you took the bull by the horns, so to speak, and introduced your beloved to her. 'Very sweet, but I fear she's too young, Timothy.' Etc. Etc.

Walking along the Hauptstrasse, in Heidelberg, turning right across the tramlines on the busy Sofienstrasse, still under my umbrella, with my computer in my backpack, on my way back to the Hotel Panorama, it seems to me that these various life events might have predisposed me to be interested in a theory of consciousness and perception that tends to give credit to the senses, or rather to experience. Had the doctors always been right in my case, had my marriage suddenly fallen in line with what marriages are supposed to be, had received wisdom, in short, tended to match reality as I experienced it, perhaps I wouldn't have given Riccardo the time of day.

Then pushing through the rotating door into the lobby, I see Eleonora about to step into the lift and call to her to wait. 'Ele,' I call quite loudly, 'hang on!' But apparently she doesn't hear. She steps into the lift and the door closes. At which point I realise that another reason why I got so interested in the Spread Mind theory is what happened to my partner last January.

Corrections

Imagine this scene. It is a January morning. You are both working hard, sitting across a table, you writing, she translating. Towards 10.30, feeling you need a break, you go down into the street and walk five minutes to a café where you both drink a cappuccino and read the newspaper together. You walk home, climb the stairs, unlock the door of your apartment. Just as your partner is about to start work again, she says something is wrong. She grabs at the table and sits heavily. My ear, she says. It feels weird. As if it was blocked. Or full of water. She stands up to go the bathroom for a cotton bud, but almost keels over and has to sit again.

Immediately you consult the Internet. You find endless descriptions of blocked ears and vertigo, any number of simple remedies. Hold your nose, take a deep breath and make as if to push air out through the nostrils. Warm some olive oil and put a few drops in the ear. Put some alcohol and apple cider vinegar in your ear, then keep it in there with a wad of cotton wool. Hydrogen peroxide, hot compresses, steam, nasal sprays, decongestants, antihistamines. The more remedies there are, of course, the less likely it is that any of them will work. Nobody has the perfect solution.

Now she has a ringing sound in the ear and the dizziness is even more severe. The NHS site says to consult a doctor after three to five days. Perhaps it will pass. With a good night's sleep, for example. You go on working through the day. She feels normal while sitting down, but every time she gets up, she has to fight to control dizziness. During the night, she keeps waking. There's a pressure in her ear. Going to the bathroom, she has to put a hand on the wall. The following morning, you insist it's time for the doctor.

So it began. Needless to say, the doctor received only on appointment and was already fully booked. Needless to say she eventually agreed to see Eleonora anyway, after a long wait. But what a relief that consultation was. It was nothing, the doctor said. It was a blocked sinus, a change in the weather, catarrh in the ear, that kind of thing. It seems one's sinuses are always blocked in Milan. The air is filthy. The doctor prescribed an anti-vertigo drug. It would all clear up in a few days, she said. Reassured, Eleonora didn't take the drug. She'd rather accept the dizziness, if it was just a passing niggle. And the dizziness did pass, towards evening of the second day. But the fullness in the ear didn't. Or the ringing sounds.

Only now that the vertigo had relented did we realise she wasn't hearing well. When I was on her right side, she didn't hear me. Almost at once, we took to arranging ourselves so that I was on her left. But it was clearly some kind of winter thing. We both had blocked noses and stuffy colds. It would pass.

It didn't pass. Five days later, Eleonora went to see a specialist, at a private clinic, for speed. An otolaryngologist. This woman confirmed that it was just a problem of catarrh in the ear and prescribed accordingly, drops, sprays, inhalations, antibiotics. Quite an array of medicines. Only on a second visit two weeks later did the doctor run a proper audiometry exam and begin to appreciate the seriousness of the problem. There was dramatic hearing loss in the right ear. Almost total. Now the prescriptions involved cortisone and pentoxifylline, a drug that would improve microcirculation in the brain and the inner ear. This was to be given daily, by injection, so we had a doctor friend come round and teach me how to jab a syringe into the upper buttock. Illness increases intimacy. I had never given injections before. I had never looked at a buttock in quite this way. If there was one thing that had worried me, when Eleonora and I first took up together, it was the idea that one day I would become infirm and she would have to do things like give me injections in the buttocks. Instead I was giving the injections. I breathed deeply to steady my hand before pressing the needle home. There were twelve phials in all. Not cheap. They had no effect whatsoever. The muffled feeling in the ear continued. The maddening tinnitus. The loss of hearing. At the next visit, the doctor prescribed an NMR brain scan. For possible tumour. But at this point, after long consultations with friends and friends of friends, a super expert had been found, one of Milan's top audiologists. So it was that almost five weeks after the onset of the problem, Eleonora finally got a diagnosis that would stick: sudden deafness.

Why had this eloquently named condition not appeared on Google when we first started typing in the symptoms? Because the most immediately frightening symptoms – dizziness, a feeling of fullness in the ear, strange ringing sounds – pulled up a whole range of common conditions, so that the rarer, but by no means unusual sudden deafness was not on the first page of hits. You pretty much have to type in 'sudden deafness' to get sudden deafness to come up first. And this is unfortunate because, as it turns out, rapid if not immediate diagnosis is crucial if there is to be any recovery. The American National Institute on Deafness explains:

> Sudden sensorineural hearing loss (SSHL), commonly known as sudden deafness, occurs as an unexplained, rapid loss of hearing – usually in one ear – either at once or over several days. It should be considered a medical emergency. Anyone who experiences SSHL should visit a doctor immediately. Sometimes, people with SSHL put off seeing a doctor because they think their hearing loss is due to allergies, a sinus infection, earwax plugging the ear canal, or other common conditions. However, delaying SHHL diagnosis and treatment may decrease the effectiveness of treatment.

In short, if you get large intravenous doses of cortisone inside forty-eight hours of onset, together perhaps with regular time in a hyperbaric chamber (breathing high levels of oxygen in pressurised conditions) you have some chance of recovering some hearing. If not, you have lost the hearing in your ear and that is that. Worse, you may have tinnitus, intermittently or permanently. Nobody knows what causes the condition. Perhaps a virus, perhaps a tiny thrombosis. Nobody knows why cortisone or the hyperbaric chamber sometimes improves it. The science isn't there. The expert audiologist began to talk about cochlear implants and complex hearing aids. She gave Eleonora a long questionnaire, which included the question: have you considered taking your life in response to these symptoms?

The following weekend, we went for a walk in a large park to the west of Milan. We needed to get out. It was a fine late February day, bright, cool and breezy, breezier than we had realised in the protected streets of the town. After half an hour, it began to feel seriously cold. The wind was nagging about our ears. To the north you could see the snow on the Alps. Walking beside her along avenues of plane trees, I was struck by the fact that Eleonora now moved in a different auditory world from the one I knew. Intelligible sounds that clearly came to me from the left or the right, all came to her from the left, while on the right she had an array of unintelligible sounds – whines and rustlings and ringing – that seemed to have nothing to do with reality. These disturbing acoustic invasions were worse when the surrounding environment was noisy: a crowded restaurant, for example, or a busy street. Or in a high wind.

Suddenly she said: 'My other ear is blocked.'

I couldn't believe it.

'I'm not hearing properly in the other ear.'

We fled. The audiologist had assured us that sudden hearing

loss affected only one ear. But we were no longer reassured by doctors. Hadn't the first two misdiagnosed her condition? Suddenly, we both foresaw a future of complete deafness. The loss of our daily chatter. Our jokes and fun. Or worse still a chaos of meaningless noises. Eleonora would be alone in her head. With silence or bedlam. Conversation impossible. Phone calls impossible. No more piano. No more singing.

We turned on our heels and almost ran to the car. Eleonora had her scarf wrapped round her face now. I drove with urgency. Back at the flat, though it was mid-afternoon, we went straight to bed and held each other tightly under the quilt. Only an hour later did we begin to relax. It was a false alarm. One of those alterations in hearing you get with a pressure change. In a plane or in high wind. We both felt exhausted.

It was around this time that I came across an article online about a man without eyes who could see. This was the *Washington Post*. A trustworthy paper. 'What does it mean to see?' the journalist asked. 'If you have normal eyesight, you probably think of sight as the ability to take a perfect picture of the world in front of you using your eyes. But that's not how vision works. The sights we "see" are produced inside our own brains. Our eyes are certainly well-designed input devices, giving us sharp, colorful outputs of the world. But what if another part of your body collected the data used to create those inner pictures instead?'

The other part of the body was Daniel Kish's tongue. 'He clicks it against the roof of his mouth and uses the echo feedback he hears to produce a kind of sonic representation of his world.'

The journalist was very excited about this. It showed that the brain could use any kind of input to produce visual representations of the external world. It wasn't dependent on the retina.

'I definitely would say that I experience images, that I have images,' said Kish. 'They are images of spatial character and depth that have a lot of the same qualities that a person who sees would see.'

Brain scans run on Kish (who had lost his eyes to retinal cancer aged three) showed that the parts of the brain that are active when one sees were indeed active as he moved around using his echo-location technique.

'Maybe you don't need eyes to see,' the journalist concluded.

I thought a lot about this article and about Eleonora's hearing loss. Essentially, what was denied was Riccardo's seamless oneness of object and experience, the idea that seeing is the consequence of an object in light encountering the body's perceptive system. On the contrary, in this view 'input' of various kinds comes from a given reality without and feeds the sovereign and separate brain with the material it needs to manufacture 'images' of that given reality within. You don't experience the world, you see 'images', or rather 'have images'. But was Kish's echo-location anything like my seeing? There was space and depth, he said. But was there colour and detail, focus and periphery? And if there wasn't, why call it seeing rather than echo-location, why insist that one way or another the brain comes up with the same experience?

Surely what was happening here was that entirely

different perceptive tools were connecting with a different set of events in the world, giving rise to a different experience of space. Perhaps it's not unlike the difference between feeling inside a bottom drawer with your hands, in search of some object or other, then, when you can't find it, crouching down to actually see inside. Both touch and sight report the same spatial situation in the drawer. In that sense they overlap. But in all other respects the experiences are quite different.

Can we say of these different experiences – Kish's echolocation and our seeing – that one is more real or more accurate than another? I don't think so. Kish moves in the real world just as we do. It's just that his perceptive equipment reacts to different aspects of it. Interestingly, it doesn't seem you can have both realities – echo-location and sight – at the same time. The nature of experience is predicated on the perceptive system. The world manifests itself differently according as you have eyes to see or a tongue to click. Or a hand to thrust into a drawer.

And hearing without ears? Eleonora's audiologist had once again spoken of a cochlear implant. The private clinic where this renowned doctor saw her patients supplied and surgically inserted these devices. Essentially, a small microphone placed behind the ear picks up sound and delivers it along wires passing through the skull direct to another device that transmits stimulation to the cochlea where tiny hairs in a spiral of liquid respond to different vibrations. Long research on the net suggested that in return for tens of thousands of euros and a tricky surgical operation, you might eventually recover hearing equivalent to a scratchy old vinyl record. Of course your brain would do its best to 'correct' for the poor quality of this auditory contact with the world, giving precedence to sounds judged important and seeking to ignore the background buzz. But this wouldn't amount to having good hearing.

This idea of 'correction' is fascinating. In the first days after she lost all hearing in her right ear, Eleonora was extremely confused about where noises and voices were coming from. You would call to her from the right and she would turn to the left, because that was where she was connecting with the sound. Any neuroscience textbook will explain how this works. The auditory nerve fibres travel from each cochlea to something called the cochlear nucleus. This is still on the same side of the head. But from here another nerve then crosses over to the 'superior olivary nucleus' on the other side of the head, the left crossing to right and right to left. Not altogether, though. Each olivary nucleus has some contact with the ear on the same side, as well as receiving the main nerve fibres from the other side. So, on each side of the head there is a point where the varying intensities of sound coming from right and left can be compared, presumably allowing an awareness of where sound is coming from. Having lost the hearing on her right, Eleonora naturally interpreted everything as coming from her left. But very soon she grew wise to this. Very soon, particularly when she was in a room she knew, the exact intensity of a voice might allow her to assume that it was not coming from the left at all. The nature of the sound experience hadn't changed from the first days, but her interpretation of it had, and with it her feelings in relation to the experience.

She was no longer so disoriented; she felt a little more confident.

This process of interpretation, or sometimes misinterpretation, which has much to do with the expectations we bring to any given event, is one reason we might get the impression that there is a mismatch between perception and reality, and hence that experience is happening in our heads rather than out there in the world. Example: driving up to Heidelberg from Milan I had a frightening experience in heavy traffic on the approach to the Swiss border. We were moving very slowly uphill, pretty much bumper to bumper, with a solid wall of container trucks stationary in the slow lane on our right. This situation went on for quite a while. We knew there were border checks because of the immigrant crisis and presumed that the police were stopping the trucks at the border to open their containers. Suddenly, I had the impression our car was slipping back and that we were about to crash into the vehicle behind. Damn! Out of nowhere, an emergency. We were rolling back down the hill. I hit the brake hard. But rather than stopping the car, this only made it slip back even faster. How could that be? Complete bewilderment. Only after another second or so did I realise that in fact the trucks in the right lane had started moving quite quickly, and since they were all huge and I could see nothing beyond them, nothing relative to which I might see they were moving, I had had the impression that they remained still and I was slipping back. As soon as I realised this, all was OK again. Except the guy behind was honking because I had braked for no reason.

Throughout this episode, what I was actually seeing – the

object, or assembly of objects, that constituted my visual experience – remained the same. There was no lapse then recovery of my visual powers, nor did I invent or create anything subjectively. I simply misunderstood my experience, in exactly the same way you 'misunderstand' when you have the impression that the static sun is falling beyond the horizon. For, actually, I was going backward *in relation to the trucks.* You get the same effect of course, but less frighteningly, sitting in a train in a station and looking out at another train stopped beside you. All at once you have the impression of moving. Then you realise, perhaps glancing to the other side at the platform, that in fact it is the other train that is moving. The only thing this tells us is that we don't perceive movement in any absolute way; what we register is the changing position of one object in relation to another, or to ourselves. As always, the perceptive system picks out a selection from reality, then we interpret, or misinterpret accordingly.

As Riccardo sees it, all illusion is of this nature: a misinterpretation. You see a shimmer in the desert and imagine water where there is no water. But you have indeed seen a shimmer, the effect of air meeting a superheated surface. And since, at a distance, eyes and brain have no way of testing whether a surface is wet, you have become used to interpreting a certain kind of shimmering surface as water. What the mirage tells us, then, is not that experience is subjective, a con job in the head, but that when we recognise water we are actually picking out only one feature of water and one that can also occur in other circumstances without water. Spend a while in the desert and you soon stop seeing water, as Eleonora soon stopped imagining that all sounds were on her left.

Riccardo came round to dinner. He had only met Eleonora once or twice at that time but was genuinely upset for her. So, she challenged him: if experience is not produced in the brain, but is identical with an external object, where was the external object that was her tinnitus, this very inner and very disturbing experience of bizarre sounds?

Riccardo has done a lot of research on phantom limbs, people who still have the sensation of possessing a hand or foot that was in fact amputated twenty years ago. And likewise on the claims of congenitally blind and deaf people to see images and hear sounds. His position on these phenomena is simple enough. In certain conditions, where our focused perception of the immediate world lapses, an old experience can return, or rather, the experience can happen again. One is still in touch, across a time delay (like looking at a distant star) with the original object/experience. This would be the case with the phantom limb. On the other hand, where there has never been, say, an experience of colour due to congenital blindness, then there can never be one. Sifting through literature on blind people experiencing colour, he established that most of those making these claims were not in fact congenitally blind and, where they were, careful interviewing showed they had no understanding of what was meant by colour. People had talked to them about colour. They were aware that sighted people distinguished objects by colour. They knew that water was shown as blue and grass as green, but they had no real sense what these terms meant.

But what about the tinnitus? Eleonora explains that she constantly hears a rustling noise, or background hum, and occasionally a distant hammering sound. Or when there's a noisy crowd – if we go to the pub or to a busy restaurant – she gets a booming noise. Oddly enough she has the same problem in the kitchen corner of our small flat, particularly when standing by the oven which has a big extractor hood above it. The room booms.

So, there are two possibilities, Riccardo suggests, which could be operating singly or in combination. One is the notion of the old experience repeating. This is more likely when you lose your hearing after exposure to a loud noise. The last noise the ear heard simply continues. But Eleonora did not lose her hearing in response to a noise.

'And the second explanation?'

That the ear is still active but not in a standard way, so that it selects a different range of sounds from the world. Clearly something is going on when she hears booming noises in a crowded room. The booming is a reality in the world, but one only perceived by an altered ear. The way, say, a dog's ears will pick up high-pitched sounds that we don't pick up. A different world from the world we perceive. So Eleonora's ear carves out a different reality, a less attractive and less useful reality maybe, because the ear has changed.

Eleonora mentioned a different explanation that she had read on the Net, that the ear itself, being damaged, produces a disturbance. Like the way sometimes you see spots in the eyes.

'No, no, no! The ear can't produce a sound by itself,'

Riccardo is adamant. There has to be an external object to have an experience. 'When you see spots in the eyes it's because there are lumps of protein in the vitreous gel in the eye. There's something there. The eye doesn't produce images that are not there. How can the ear produce a sound? It can only transmit, one way or another, the vibrations it receives.'

I'm always surprised by his vehemence, his extraordinary conviction.

'Think of the way,' he says, 'they invite children to cover an ear with a seashell and hear the sea. It's something like that. You change the reality outside the ear, around the ear and the sound changes. In any event, it's always a reality that's out there, not in your head, otherwise you wouldn't need to go to a crowded pub to hear it. It's like the way a short-sighted eye picks out a different visual reality than a long-sighted eye. But both realities are just as real. There is no absolute reality to be experienced, or no absolute perceptive system to pick it out. Only different perceptive systems each making possible different manifestations of the world. A cat sees one thing, a bird another, a person with sunglasses another, an X-ray machine another, and so on. Who could say visually what an object absolutely is? Kant's *Ding an sich*? We see parts of things. And the same goes for sound.'

'Example!'

With his usual self-ironising flourish Riccardo opens his laptop. He has stumbled upon a revealing case, he tells us, and invented a little experiment, to do with colour after-images. He fires up PowerPoint, shows us a bright red slide and asks us to look at it intently for twenty seconds. Then switches to another slide. What colour is it? It starts by looking a bluish green, but then fades to a light grey. Now there is another slide with two colours side by side, one green, one bluish green, or as the label beneath says, 'Cyan'.

'Which of the colours did you see after the red slide? The green or the cyan?'

We both agree it was the cyan.

'So, when you look at a bright colour for a while, your perceptive system gets used to it and begins to filter that colour out, probably the same way Eleonora is trying to filter out her tinnitus. A form of correction, if you like. The result is that when you turn away from the colour you're blind to it for a short while. In the philosophical and neuro-science literature the brain then produces an opposite colour – in this case green, they say – that is not really out there. And this is supposed to prove that experience is in the head. Colours are in the head. They say green, by the way, because they're using Hering's theory of colour opponency. They're deducing green *from a theory*. In fact, when you run this test on people, what they really see, like you did, is cyan, a blend of green and blue, which are the two colours that, together with red, go to make up grey, or white. So when you're red blind for a moment and look at white or grey, you don't invent the cyan; you see the colour that is always there but normally hidden by the addition of red that turns it into white or grey. What you're seeing is grey minus red.

'Try this.'

He shows us a red slide for twenty seconds, then another.

This time it looks green at first, then gradually settles into a bright canary yellow.

'Yellow minus red is green. Right? You were seeing the green in the yellow that then is hidden again as the red comes back.'

The moral?

There are all kinds of 'correcting' reactions which alter perception for a moment. But the colour you see *is* always a colour that's really there, even if it requires a temporarily altered perceptive system to see it.

'Try this,' Riccardo says.

Again he shows us the red slide, but now he stands up and turns the light off. The shutters are closed and the room is dark. After half a minute he snaps the laptop shut. The red is gone.

'So, what after-image do you see?'

The room is pitch-black. We can't even see each other.

'Nothing at all.'

'Right. Because there is no light. Everything's black. Take red from black and you still have black. If the brain was producing the colour, it surely wouldn't be worried by that.'

The nice thing, Eleonora remarked when, towards midnight, Riccardo had set off to see one of his lady friends, was that during our intense conversation she had stopped noticing the tinnitus.

A couple of months later, invited to Boston to give a paper at a conference on translation, I persuaded Eleonora to come along so that she could book an appointment with the Massachusetts Eye and Ear Hospital and see one of the world's leading experts on sudden deafness. Results of the now endless tests and scans she had been through had been sent on ahead, together with a detailed account of the onset and progress, as it were, of her condition. After various preliminaries with secretaries and nurses, the great man himself finally arrived and sat down beside Eleonora to discuss her case. He had her dossier under his arm and began to leaf through it. All the tests had been correctly carried out, he said, and were all of excellent quality. Leaving aside the unfortunate delay in reaching a diagnosis, all the subsequent prescriptions and actions taken were the best that could have been done in the circumstances and exactly in line with what he would have done himself. He was impressed.

Then he looked up and said:

'There is nothing you can do for this ear, Miss Gallitelli. It will never hear again. So the best thing is to do nothing. If you continue to consult doctors hoping for a cure, you will be wasting time and money, and making yourself anxious into the bargain. Do not be tempted to use a hearing aid or, worse still, to have an aid implanted surgically, since this would only distort the excellent sound you are getting from the other ear. Take comfort from the fact that the tinnitus has mostly subsided. Be thankful that your good ear works well. Be careful with it. Don't expose it to loud noises; don't take up scuba diving. In general, do your best to forget the whole thing, since, I repeat, all case histories show that people with the symptoms you have do not recover any hearing in the damaged ear.'

Eleonora was extremely upset by this interview. The world she had known, when she sat at the piano and was flooded with sound on all sides, was now officially gone. Aged twenty-nine, with no warning and no explanation. For a day or two she was deeply melancholy. Then the frankness of this doctor's advice began to produce its positive effects. Onward. Experience had changed, but brain and body would be working hard to correct for it, to forget that life had ever been different. And indeed five months later she has already largely forgotten. So that when, back in our room in the Heidelberg hotel I tell her that I called to her in the lobby and she didn't hear, all she does is smile and embrace me and invite me to bed for an afternoon nap.

Later, before looking at my notes for the interview with philosopher Thomas Fuchs, I asked Google whether anyone had tried to map out what goes on in the brain during love-making. Surprisingly, the search came up blank. Nobody has done this, or not that I could find. There was a paper, however, by Semir Zeki, our famous neurologist at University College London, entitled 'The Neural Basis of Romantic Love' in which he sets out, he says, 'to chart the brain areas responsible for producing this emotional state'.

To think I had been so naive as to imagine that Eleonora was involved . . .

Loops

'What about dreams?'

The interview with Thomas Fuchs could hardly be more different from that with Sabina Pauen earlier in the day. We're in the Psychiatrische Universitätsklinik, Heidelberg's psychiatric hospital, where Fuchs is the Karl Jaspers Professor for the Philosophical Foundations of Psychiatry and Psychotherapy. Born in 1883, Karl Jaspers was a German psychologist and psychiatrist whose work on mental illness led him to move into philosophy. Fuchs spans the same fields. He works as a psychiatrist looking after schizophrenic and depressive patients in the psychiatric hospital, and at the same time writing about the nature of experience and being in a way that brings together medicine and philosophy.

However, what's really different is that this time *I'm* the one asking the dreaded question for all models of consciousness: 'What about dreams?'

'Ah, the argument from *dreams*,' says Fuchs knowingly. 'Generally used to demonstrate that all experience must be created by the brain, inside the head. Isn't that right?'

Fuchs has a very lively manner that I'm finding hard to

pin down. He's intellectual but mischievous with it, and somehow slightly proper and crusty too. He has a long sharp nose and rimless glasses, but his body is solid, energetic and apparently comfortable, if a little restless. He keeps shifting position, moving his hands. His clothes are half formal – a smart jacket, and shirt, good trousers – and half casual – no tie, nothing remotely 'ironed'. He is welcoming, makes a cup of tea for me, produces some fancy biscuits from a box, but he's also very slightly distant, perhaps a little cautious, and the biscuits are stale. They don't crunch. Presumably he keeps them in his office for guests, but doesn't have guests very often. In a similar situation I would have eaten them myself ages ago.

I say 'we're in the Psychiatrische Universitätsklinik', using the present tense, but in fact there is something I must confess to the reader at this point. The truth is I began this book a good six months *after* the events I describe, and though I took notes during and immediately after the interviews, I did not record these conversations. I had not thought I would be using them for any more than the relatively short article commissioned by the Deutsch-Amerikanisches Institut. Also I've always suspected that putting a recorder in front of people changes what gets said. As a result, what you're reading here is my reconstruction of these meetings, based on a rereading of all the material I had from the professors I visited, my notes of course, and then a huge effort of memory, an attempt to relive these moments as far as is possible. It's interesting in this regard, that what I remember most easily of all three interviews, and likewise the various meetings with the folks from the

Deutsch-Amerikanisches Institut, not to mention the many coffees, drinks and meals in bars and restaurants with Eleonora, is my position in the room. Or more precisely my position in relation to the person I was with. So in this case, for example, I recall that Fuchs, having first come out of his office and asked me, in the corridor, to wait a few minutes while he finished a tutorial with a student, then invited me into his room and offered me a choice: either the two leather armchairs by the window or face to face across his big desk. I chose the across-the-desk situation, with the window behind me as, being more serious, more likely to produce a focused conversation, though later we did move to the armchairs for tea. The quality of the biscuits I also recall very clearly, though I made no note of this of course. That is, I recall, six months on, my disappointment when I picked up a rather fancy Viennese biscuit, a pale elongated oval tipped each end with dark chocolate, and bit into it, finding it slightly soggy. Apparently a part of me was more concerned about the quality of the biscuits than the content of the interview.

But what I most remember about this meeting with Thomas Fuchs, as I feel back towards that afternoon of 22 September 2015, a day lived with great intensity but now maddeningly elusive, was its curious mood of mixed pleasure and anxiety. And the more I think about this the more it seems it had to do with a confusion about age and hierarchy. Fuchs was born in 1958, I in 1954. So I'm older than him. Yet throughout the interview I behaved as if he were older than me. You could say that this is because he was the authority to whom I was appealing, the philosopher, the psychiatrist, the expert. He was on the business side of the desk, with shelves of forbidding tomes behind him, I on the suppliant side, where no doubt his student had just been sitting. But you might also wonder if this odd behaviour of mine wasn't partly the consequence of my being the youngest of three children, someone who grew up aware that all the key players in his life were older. I remember I was in my mid-twenties before it dawned on me that there might now be people who would relate to me as older than them, and when that did occur I didn't know how to respond; I hadn't learned how to be the older person. Even today I will catch myself treating younger people as older. It's my default setting. Which makes no sense at all in one's sixties. And maybe some of the ambiguousness in Fuchs's manner was his difficulty in reacting to my unexpected deference. In any event, this slight unease manifested itself throughout the interview, on both sides of the desk, as a sort of repressed competitiveness. Fuchs wanted to display his competence, as though giving a lecture to students, an attitude I no doubt invited, while for my part I was eager to show I was no fool; I had done my homework.

But enough digression! I hear the reader object. Forget your mood, Parks – who cares? – and get on with the substance! What was said at this meeting, and why were you asking Fuchs about dreams rather than he you? But there is no digression here, I promise. All this background is appropriate and necessary. Because Fuchs is a phenomenologist, one who believes that conscious states are best understood through the meticulous descriptions of those who experience them, an approach he inherits from, among others, his great predecessor Karl Jaspers; only the subject – me in this case – can express his subjective experience and we must give credit to what he tells us, even if we have no access to it. Then when it comes to the how and why of consciousness, what it is made up of, where it might be located and what purpose it serves, Fuchs is what they call an 'enactivist'.

Let's not be put off by the fancy term. Enactivism is an attractive approach, one that I'm constantly tempted to fall into and would very likely have adopted, if I didn't have Riccardo constantly and rather sternly reminding me of its shortcomings. Essentially, the enactivist – and Alva Noë, the philosopher who dared to criticise *Inside Out*, is one of the pioneers – believes that consciousness arises from our purposeful dealings with the environment, starting with the hungry baby blindly reaching out for something that will satisfy it and consequently finding mother's breast, finding, that is, the part of the world that meets its need. In this version of events, the environment is not passively perceived by our eyes, ears, nose, etc., but actively *constructed*, from birth, through our interaction with it, as we learn to create a useful fit between our requirements and what's available. Same with our social relations. Our consciousness of other people is constructed through our dealings with them, our fitting in with them. So every new act of perception, whether it be seeing, hearing, tasting, touching or smelling, is built on all previous similar acts and everything we learned from them, meaning all the ways those experiences changed our highly plastic brains. My consciousness of Fuchs today, his of me, will be conditioned by all the habits we separately bring to our meeting, and then our explorations, as it were, of each other, through our conversation. Crucially, there is no experience, no knowledge, without the presence of the object, or person, with whom we are interacting. Or put the other way round, as Fuchs does in one essay, 'Objects exist only in so far as they are objects of action.' Consequently, my mind is not trapped inside the skull, producing mental representations of a reality that might actually be quite different from what I imagine; rather the mind is present, 'extended' they say, throughout the body and even into the environment as the two connect and interact.

You can see there are some similarities here with Riccardo's Spread Mind theory. Above all, enactivism rejects the idea that we only need to hang on for the neuroscientists to perfect their knowledge of the brain for the problem of consciousness to be solved. And in fact, no sooner am I sitting down than Fuchs asks me who else I'm seeing in Heidelberg and when I tell him Sabina Pauen, he nods and says he knows something of her work and approves, but when I tell him Hannah Monyer, he laughs and says, 'Ah, rats and neurons, synapses and ganglions! The brain as the unique seat of consciousness!'

'It's the dominant position,' I point out. 'The neuroscientists have made extraordinary progress.'

'With their clever machines!' The philosopher is expansive and charmingly scathing. 'They're never going to find thoughts and perceptions in the brain, only the neural activity that supports them. But because they study the brain, they take it to be the only thing that matters.'

Fuch's voice is pleasantly deep. His accent has the

distinctive sound of someone sent to an upmarket language school in the London of the 1970s or 80s. Slightly plummy, so that you have to be constantly reminding yourself that you are not talking to a member of the English upper classes. On the other hand, the decor, in this nineteenth-century building, all polished wood and dark upholstery, is reminiscent of the kind of college environment where I regularly heard this sort of accent in the mouths of Cambridge professors back in the 1970s. So if, as enactivists believe, knowing an object amounts to knowing how to use it – because that's how you came to consciousness of it in the first place – and knowing another person is knowing how to interact with them, then clearly things can get muddled when certain signs tell you that you have dealt with this kind of person before, you know them, though actually this is not the case. Fuchs *isn't* a Cambridge professor in the 1970s; he's a German psychiatrist and philosopher, now in 2015; nevertheless I feel I'm back there as his student.

He takes over the conversation with what is obviously a spiel he has performed a hundred times. Our consciousness of the world, he says, develops round a tension towards action. 'Looking at the door, for example' – he turns to his right where a handsome wood-panelled door leads back into the corridor – 'I'm aware that I can reach out for the handle and go through it. I'm aware of my body's relation to the door.' He talks for a while about this, and, thinking back on it now, it occurs to me that this approach fits in with the dominance of physical position in my memories. When I think of any major event in my life, my first meeting with my wife, the birth of our first child, my mother's death, what I remember is where the people involved sat or stood or lay in relation to me. Perhaps readers might like to pause for a minute or two and experiment with that. Is it the same for you?

'Your eye focuses on the handle,' Fuchs says, 'which is your point of interaction with the door. Your awareness of the door is a knowledge of what it would mean to turn the handle and open it. We call that affordance. The handle affords you the possibility of turning it, of engaging with it. It has a well-defined potential. And that knowledge is in your hand and arm, your body, as much as your brain. Same with any tool, or knife, or pen. Your sense of the affordance they offer, your knowledge of them, has been constructed using these things. You don't passively see them, you are tensed towards engaging with them.'

'Remind me of the brain's role in this,' I ask. Actually, having read the various articles he sent me and listened to a couple of lectures of his on YouTube I have a pretty good idea of what he will say. And, lo and behold, he says it, sometimes almost word for word. But it's always interesting to hear people present their views orally, to get a sense of their confidence, or otherwise, in the ideas they have chosen to attach their names to. The fact is none of us can hold a complex theory whole and simultaneous in our heads. Whenever it's brought to the surface you have to reconstruct it, perform it, one thing leading to another, like a child repeating a poem they have learned. And often it's during the performance that you discover something new, or realise that something doesn't quite work. There's a missing link, or the road forks invitingly where you thought it offered no choice.

There are two basic aspects of experience for any creature, Fuchs says. A continuous bodily background feeling, your mood, the presence of a physical self, your attunement, or otherwise, with your environment. '*Befinden* we call that in German. Or simply *Leben*, if you like. The feeling of being alive. And then there is drive, instinct towards contact with the environment, towards change. We can call that *Erleben*, living through. Or conation, from the Latin *conatus*.'

It's not clear what the foreign words really add to these ideas, but they do give Fuchs's presentation a nice sort of sonority.

Obviously, Fuchs says, the drive is related to the underlying mood. If you feel hungry, you move to eat. If you feel unhappy, you try to change your situation. What the brain is doing is facilitating and mediating this constant process of attunement with the environment, involvement with the environment. So whenever you act, when you turn a handle, say, and open a door for the first time in your life, you lay down a pattern of neural activity in the brain which is potentially reactivated every time you come in visual contact with a handle and then actually activated when your hand grasps the handle.

You could say that at that moment a kind of open loop closes into a successful circuit. The loop is this neural pattern in the brain, but when it closes it also includes your body and the handle itself. The whole thing, the lived action, is mind, thousands upon thousands of neural patterns engaging and re-engaging the world in all kinds of ways. So the mind is not produced by or in the brain; rather the brain is formed and constantly modified by the mind, which is your embodied experience in the world.

All this is in line with the papers Fuchs emailed me some weeks ago, and has obvious connections with the work Sabina Pauen is doing. The child's brain changes when it sees a face or plays a game with tools, so that when the face reappears or the game is proposed again, the brain recognises it, connects with the past, even though the face isn't 'stored' in the brain. It's not entirely unlike the idea we looked at before, that the brain constructs a sort of neural lock for the 'key' of a specific object or person. When object or person turn up again, the lock turns, you have recognition.

Still, there is something curious about this idea of enactive loops. Riccardo's old theory of experience as process, a unity composed of perceptive system, object and everything fizzing between, was materialistic and pragmatic. Everything could be accounted for in physical terms: the object on one side, the neurons on the other and between them the photons and the sound waves and the molecules that give us smells. The whole ensemble was experience. Enactivism requires something else, drive: experience can only happen thanks to our constant tension towards engagement with the environment. But are we really constantly tensed like that in every act of perception?

'Let's go back to the door,' I suggest. 'I understand what you're saying and it makes perfect sense to me, that I have this bodily knowledge of doors that isn't just located in my brain, so that when I open a door there's an experience made up of handle, hand, eye, body and brain. On the other hand, that doesn't really account for the amazing detail I get when I look at the door: the graining in the wood, the scratches, the old-fashioned panelling. It's not just any door. And then all round the door, the lintel, the walls, yellowing paint, the ceiling, I'm not in a habit of engaging with ceilings. But I see it in exactly the same detail as I see the things I don't engage with. So if an object exists only in so far as it is an object of action, why does the ceiling exist?'

Fuchs laughs. The perceptive system that developed out of our exploration of the world prepares us for possible engagement with, or possible flight from, all phenomena. And often, of course, detail is essential for the way you engage with it, so the system that developed over the evolutionary process allowed for detail even where it wasn't essential. This reflection reminds me of a long paper by J. Kevin O'Regan and Alva Noë which tries to describe the enactivist position in neurological terms. One of the curiosities of the brain is that though there are evidently areas specific to different kinds of perception, nevertheless the neurons in these areas don't seem to differ from each other in any way, so that you have to wonder how they know that what's exciting them is a sound rather than a smell, a sight rather than a taste. O'Regan's and Noë's long and difficult paper – 'A Sensorimotor Account of Vision and Visual Consciousness' – offers a hypothesis about how the exploratory urge, our atavistic drive, originally meshed with the different patternings of visual and auditory reality as experienced through eye and ear to create perceptual experience. It's an extremely elaborate hypothesis that could only be verified, if at all, by the most sophisticated neurological experimenting.

In any event, since I'm not equipped to argue against O'Regan's and Noë's theory, I now have just one last piece of ammunition with which to ruffle Fuchs's feathers, or get him to tell me something I haven't heard.

'What about dreams?'

If conscious experience is the laying down and then activation of loops, neural patterns that lock into external phenomena, what is a dream? How can the loop be completed and the neural pattern activated when the eyes are closed, the body asleep, and there is no world to engage with? Surely here experience is simply generated by the brain? And if it can generate dream experience, why can't it generate waking experience too, with a little help from various 'inputs'?

Fuchs isn't ruffled at all, and as he speaks I realise that actually he is not profoundly interested in this kind of debate. What matters to him, in the end, is the relationship between his view of how experience works and his patients. He is attached to the enactivist approach because it offers an explanation of what is going wrong in their lives and suggests ways of helping them. Clearly, he doesn't believe these ideas can be proved wrong or right at a neurological level.

'First, dreams are not like normal visual experience,' he says. 'Or normal experience at all.'

Actually, I agree with him on this. Dreams seem very real, it's true, but there is a different quality to them. 'My dreams have a rather closed-down feel,' I tell him, 'as if there were not much space, nothing really out there around the core of the dream.'

'Right, but most of all we have no sense of agency in dreams, because there is not the enacting loop with a real world that affords active engagement. Aside from the fact that it's absurd to build a theory of consciousness around a special case like dreams, rather than ordinary daytime consciousness, to say, since I dream a door without a door being there, it means that even when the door is there it's just invented in my brain. It's waking experience we should concentrate on.'

'Still, dreams are a more or less universal experience, aren't they? And we do see doors and lots of other things in them, sometimes frighteningly clearly. Something is going on. For hundreds of years philosophers and scientists have been using them as a paradigm of the way perception is generated in the brain.'

Fuchs is patient.

'Dreams are less stable than daytime experience,' he points out, 'things keep changing, as if the brain couldn't sustain visual experience for long without contact with the real thing.'

I'm not sure about that. Odd though they are, some of my dreams do seem disturbingly consistent, especially the worst ones, which usually have to do with my having killed someone and wondering how long it will be before the police catch up with me.

'Noë, I seem to remember, suggested that dreams were stimulations of the neural patterns, the loops, but in the absence of the object in the real world that would normally complete the loop and bring about conscious experience.'

'That's right,' Fuchs nods.

'But in that case, it seems he's saying experience might be experience of the loop, or even just the neural patterns forming the brain part of the loop, not the object that closes it.

And if that's the case we're back with experience being in the head, at least potentially, while the object is separate from the experience, out there, not really necessary for what's going on in the head.'

As I use this 'argument from dreams', one that Riccardo has had levelled against him a thousand times and has spent hundreds of hours of thought and speculation building a response to, a response, as we'll see, that requires a complete revision of how we think about more or less everything, I can't help but feel its simple power. You can say what you like about dreams not being the same as ordinary experience, but they do seem to be an experience of some kind, a consciousness of some sort, and they do appear to unfold entirely in the head. The internalist camp really does have a big weapon here. I'm dreading the moment when I have to explain Riccardo's rebuttal, if only because I'm not sure that I'll be able to perform the argument from beginning to end; it is such hard work, conceptually.

'Dreams are certainly a challenge,' Fuchs acknowledges but repeats that for him they don't really count. What matters is the sphere of daytime experience and action. And here the enactivist view allows you to make sense of mental illnesses like depression and schizophrenia, because you can understand them as disturbances in the way reality is constructed through engagement with the world.

I ask for an example and he begins to talk about the experience of time in schizophrenia and depression. The schizophrenic experiences time as speeded up. Things are happening too quickly to be grasped. The depressive experiences it as slowed down. Time and timing is one of the essential elements in our interactive construction of reality. We attune ourselves to the speed things happen at, our engagement with the world when walking, running, driving. Or with people in the back and forth of conversation, of gesture and counter-gesture. This is part of the neural patterning established with every experience and influencing future similar experiences. With people, in particular, we don't represent what they have said to us in our heads and interpret it before responding, as the internalist approach would have it, we simply mesh with their body language and conversation, largely unconsciously. This kind of reciprocity, where timing is essential for interaction, breaks down with mental illness and is associated in the depressive with a general loss of drive to engage with the world, in the schizophrenic, with a paranoia that things are happening that shouldn't be and that threaten her.

Fuchs, I should say, has a habit, when both writing and speaking, of using either masculine or feminine pronouns, more or less at random, to mean 'everyman'. This is admirable, no doubt, from the point of view of political correctness, but disturbing for the flow of conversation; the fact is that though I'm used to hearing an undetermined person referred to as 'him', when I hear a 'her' I always imagine there must be a particular woman we mentioned earlier. For a moment I'm trying to grasp who he's talking about. Clearly I just don't have a loop for this particular linguistic phenomenon. I can't get used to it.

'But is this time problem the result of an anomaly in the brain, or in the world?'

'Any problem in behaviour is present in the brain, the

body and the world, because consciousness is an embodied thing regulated through the brain.'

I take his point, but there does seem to be a little bet-hedging here. I mean, should one take drugs or try to sort out the world one is living in? My experiences with depression fit in with the description Fuchs gives of the condition in the paper he sent me: 'The senses become blunt, the gaze tired and empty, the taste stale. The general decline that manifests itself in numerous organ systems also lends a void, blunt or dull colouring to the environment. With the loss of attunement, feelings of distance and unreality may arise.' But when Fuchs claims that this is 'triggered by a desyn-chronisation of the individual from his environment, which then develops into a physiological desynchronisation', this seems a shade generic and unnecessarily technical. In my case, depression was triggered by an awareness of being ab-solutely stalled in my marriage and private life. Something had to be done but I couldn't do it. No doubt alterations in the brain occurred – how could it be otherwise when one feels frustrated and paralysed for long periods of time? Cer-tainly, the body, as Fuchs describes it, was experienced as heavy and difficult, everything became difficult; I had no drive, no appetite. Even a bright spring day looked duller, deader, as if nothing could really happen. So ordinary per-ception was affected. Or, as Riccardo would say, the objects my body singled out in the world were different. Just as real, but not the same. However, the way to get over this un-happy state was by sorting out my stalled life, not taking drugs. I was offered drugs, and I took them for a while, but they only threatened to make the whole thing permanent, by making it vaguely bearable.

This unhappy memory, brushing across a few neural loops (every memory is 'enactment in a different context', enactivists believe), spurs me to launch into a rather impassioned argument. The conversation needs shaking up. Conversations are not scientific papers, they need life and action.

'Why,' I suddenly ask Thomas Fuchs, the Karl Jaspers Professor for the Philosophical Foundations of Psychiatry and Psychotherapy at the Psychiatrische Universitätsklinik in Heidelberg, Germany, 'why if, as you claim, it seems obvious that outer and inner, mental and physical, are one in the creation of conscious experience and hence mental illness, why, that being the case, do people seem to prefer the idea that such illnesses are organic, entirely located in the brain, entirely physical? That they are something to throw drugs at? And why more generally – because this is what really interests me about the whole subject – does modern society prefer the view that mind and consciousness are produced in the brain to the notion that on the contrary, mind extends into the body and even into the things we interact with? Because if, from a scientific standpoint, neither hypothesis has been proven, then the choice we make is based on inclination not facts. And most people incline to the internalist view, even when it doesn't seem the most intuitively obvious view.'

Fuchs agrees wholeheartedly with the premise of my question, which seems to cheer him up. And he has an immediate answer. Because, he says, people love a concrete, 'scientific', objective explanation, that is, an explanation based on evidence – medical tests, X-rays, scans – external to themselves, something that puts their mind at rest. And he describes the excitement and satisfaction of one bipolar patient when he was shown a brain scan that indicated an unusual pattern of brainwaves. 'Then he could say he was really physically ill. The illness had been objectified in the scan. He had seen his illness, as something separate, in a photograph, rather than just suffering it. And that separateness introduced the idea of competence and possible control. If we can photograph something, we can treat it.'

This does not sound so different from my father's response to seeing the X-rays that announced his death sentence.

Pleased with our conversation, Fuchs jumps to his feet, offers to brew more tea and again proffers the biscuits. We move to the armchairs, as if the business side of the afternoon were over. So it is between sips of fresh tea and nibbles of stale biscuits, that I hit him with the Spread Mind theory.

Sitting on the armchairs, I should say, the dynamic between us immediately changes. We are not face to face, but at three-quarter profile. And our bodies are relaxed and partially reclined, rather than upright and projected towards each other. Fuchs is more relaxed, less the professor, while I finally stop treating him as if he were older than me. In any event, I'm doing the explaining now.

He listens intently. Perhaps because I explain things badly he assumes for a moment that the idea is akin to Bishop Berkeley's position, that the world is only there in so far as there is a subject to perceive it. 'It takes an egomaniac,' Fuchs observes, 'to imagine that he is responsible for the world's presence.' I hasten to put him right. The world is there, of course, whether anyone sees it or not, but our bodies with their perceptive systems pick out only a certain subset. And that subset, the object, *is* the experience. Not the absolute object, of course, if such a thing exists, but the particular object that a human body can perceive. Your experience of the door and the handle *is* the door and the handle. Not the loop. Your experience of the sky is the sky, not the neurons in your head.

Fuchs is intrigued by this and toys with it. He agrees that to think of experience as outside the head, not locked inside, to feel and accept that we are one with the world, that the world is included in the experience, is heartening and therapeutic. And as he speaks and is surprisingly open and enthusiastic, I realise now that my perception of him during the earlier part of the conversation might have had to do with his defensiveness about his ideas, or my anticipated anxiety about how he would react to mine. Why do we all care so much about the ideas we have? Why do we feel obliged to attach our identity and self-esteem to them? In any event the English novelist and the Karl Jaspers Professor are suddenly getting along like a house on fire. In the Heidelberg psychiatric hospital. And I tell Fuchs that though I find my friend Riccardo's theory fascinating and credible, I don't think there is much chance of it substituting the presently dominant internalist position for two reasons: first because people like to think they possess their experience, it is in their heads, not in the world, in their brains, something special and separate. Second, because the internalist position, the position of most neuroscientists, by always promising that it is about to solve the problem of consciousness, but then reliably failing to do so, actually allows us to reconcile two contradictory but very powerful desires: one that the world be explained scientifically, so we have control of it, and one that consciousness remain a mystery, so that we can feel special. That is, we have preferred a 'science', a situation, a rather extravagant and expensive dead end, if you like, that spares us from deciding between those two desires. Neuroscience's special mixture of success and failure, fantastically interesting information about the brain but no convincing overall explanation, guarantees its popularity. Because people like the idea that their consciousness is special and mysterious and remains tantalisingly out of reach of even the most sophisticated machines. That's why David Chalmers was so crazily successful when he came up with the expression 'the hard problem', and talked about consciousness entailing some substance or phenomenon as yet beyond our grasp. 'It's amazing how many philosophers seem pleased with the idea that consciousness is beyond our understanding. Because it leaves man with his specialness, his desire to be beyond any explanation.'

As I hold forth, Fuchs drinks his tea and nods. 'Your friend's idea,' he says, 'like any new idea, can only succeed when the old generation die off. No change in philosophy ever really wins by argument; because no one at the pinnacle of science or philosophy, no one who has become an authority, ever really admits they're wrong. Only when the big names die do you have a chance to introduce something new.'

On this pessimistic, but also curiously satisfying note, the conversation closes, and I am halfway down the stairs before

I realise that Fuchs did not even ask me about dreams in the Spread Mind theory. Or time. And I'm tempted to run back upstairs if only to see if I'm capable of explaining it, and whether I'll feel I really believe in it as I try, but after only a moment's hesitation I drop the idea. All I have learned about interrelating with others, particularly professors who grant you a circumscribed quantity of their precious time, tells me you don't do this. And I hurry outside where some of the most time-worn loops in my old English neurons connect with thickly falling raindrops to make me conscious that I should have brought my umbrella.

Kaiserschmarren mit Apfelmus

To recap, at the end of my first Heidelberg day, we have come across or described three positions on consciousness, which we've decided to define simply as the feeling that accompanies our being alive, aware of perceptive experience.

1. The popular and orthodox view: consciousness is produced by your brain and exists exclusively in your head. This is supported by almost all neuroscientists and many philosophers. Most textbooks give this view as proved.

2. The minority enactivist view: consciousness arises from our active engagement with the world and requires both subject and object to happen, so conscious experience is extended through the body and into the environment. Available in a wide range of variations, this view is supported by some philosophers and a few neuroscientists.

3. The minority minority view, the Spread Mind, in which experience is made possible by the meeting of perceptive system and the world, but actually located at the object perceived, identical with it even; in short,

experience is the same thing as the object. Excogitated by my friend, Riccardo Manzotti, there are hints of this position in the Presocratics, in the Buddhist tradition, in William James, in Bishop Berkeley, and many, many artists, novelists and poets. Virginia Woolf, Samuel Beckett, Wallace Stevens. 'What are called outside and inside are one and the same,' Beckett wrote. 'She being part', Woolf writes of Mrs Dalloway, 'of the house there, being laid out like mist between the people she knew best . . . it spread ever so far, her life, herself.' 'This here and now seems to require us,' says Rilke. And so on. Intuitions, not science.

'So,' asks Eleonora, who loves these conversations, loves the feeling that we're trying to understand something *together*, 'in the textbook view, do they have any idea how it actually happens? In the head.'

We are in a restaurant poring over a menu entirely in German. An interesting problem for a vegetarian. The waitress has made it clear she has no English whatsoever. All around us are furnishings, inscriptions and twee paraphernalia that our limited knowledge and experience of the country can only read as generically German.

'In terms of concrete facts, they have thousands upon thousands of possible "correlates of consciousness". That is: when you have a specific experience in controlled conditions they can record certain apparently correlated things going on in certain parts of the brain, usually extremely complicated things, chemical and electrical combined, and usually going on incredibly fast and in more than one part of the brain, which turns out to have an awful lot of parts. They've produced impressive brain maps with arrows moving in all directions relating to different kinds of perceptions. This part of the brain lights up when you see a face, this part when you brush your teeth. But they don't know if or how those happenings in those places actually "produce consciousness". They have a number of hypotheses: that the conscious experience is actually identical with the neural activity, for example. But there the problem is that neural activity and experience look pretty different. How is it that this neural activity is this intense smell, and this other neural activity, which from outside to all intents and purposes looks pretty similar, is a Beethoven symphony? Another idea is that consciousness is a special quality that "supervenes" on, or simply emerges from, the neural activity, something that just starts to happen when you have enough neurons sparking and they're all connecting and interconnecting and feeding back and forth in sufficiently complicated ways. Not the neurons themselves then, but a kind of cloud lifting off them. But at that point you might as well say it's magic, because they don't know why consciousness would emerge or supervene from huge quantities of connecting information. There is no scientific principle, or even hypothesis of a principle that purports to explain such a metamorphosis. David Chalmers, who's assumed a sort of star position in the discussion, has talked about a possible substance we have yet to identify. Something there we can't see or even locate. Which sounds like magic again. Actually, the really dominant position is to say that, though you are convinced consciousness *must* be in the brain, there is nevertheless a conceptual chasm such that we can't understand how it's happening. A philosopher called McGinn said it would be a hundred years and more before they could bridge the gap between the soggy grey matter and the Technicolor phenomenology of experience.'

'Optimistic.'

'Very likely. The main camp in the end is the defeatist camp. Most people agree it's all in the head, but then they go on to say it's something beyond our present ability to track down. It's physical, but we haven't even begun to understand the real nature of the physical world, quantum physics et al. Bertrand Russell, for example, said the problem was we just didn't know enough about matter. Galen Strawson, who's a brilliant mind, still signs up to that. So it's been a long time. In fact, the neuroscience position is essentially this, that their research is directed at bridging this formidable gap between brain cells and conscious experience, except that no one seems particularly worried that this isn't going to happen anytime soon, maybe because we're not talking about curing a killer disease, just understanding who we are. And perhaps we don't really want to understand who we are because it could be a little frightening. We would feel less special. As long as there's mystery there's hope.'

'Neuroscience will get there sometime in the future.'

'The distant future. They issue promissory notes – we're making progress, we'll get there some day – so long as you keep giving us enough cash for expensive research. It's a sort of Ali Baba's cave – the exciting, always partial results – and a cul-de-sac at the same time.'

Talking about things postponed into the future, however, we still haven't ordered our meal. The waitress is hovering over us, but we haven't been giving our attention to the menu. Because we can't understand it. It might as well be quantum physics. We're going to have to order on a hit-and-miss basis. In fact, the whole evening has been miserably hit-and-miss and could well be used to illustrate the enactivist position. Feeling hungry, we set out from our hotel around eight o'clock to explore the environment and hopefully interact with some situation that would meet our need. Conation, drive. It was raining heavily. We had umbrellas, but I had no hat and it felt cold. We decided to explore the labyrinth of streets between the Hauptstrasse and the river, where, according to Google Maps, most of the town's restaurants were situated. It looked easy. And in fact, had this been Italy, or England, we would have settled on a place in no time: some facade, some look, the impression given by peering through a window, or studying a menu, would have activated neural loops laid down in the past by experiences that were more or less satisfactory. Given the meteorological conditions, we were not setting the bar very high.

But in Germany we can't read the restaurant exteriors. Some seem irretrievably gloomy. Some seem impossibly busy and noisy, which is a problem for Eleonora's ear. Tourist traps alternate with flea traps. Or that's our impression. The rain is coming in windy flurries and urges us to decide. One of the umbrellas has a broken spoke. We're now feeling distinctly hungry. My bald head is getting colder by the minute. But we can't settle on any particular place, because we want to avoid *an unpleasant experience.* Presumably there are loops that warn you that it would be a bad idea to close them without having more information.

Meantime, the Heidelberg street plan is a maze. We say 'OK, this place will do, if we can't find anything better in the next ten minutes,' but then not finding anything better in the next twenty we can't find our way back to 'this place' either. Where was it? The streets are all narrow and crooked. We can't distinguish between them. And the rain comes heavier. We poke our noses in a few places and beat a rapid retreat. If knowing a restaurant is knowing what it's like eating there, we don't know these places. One time I'm for going in, but Ele isn't. Then vice versa. She thinks this Thai place is a good bet but I don't want to eat Thai in Heidelberg. Until finally the balance tips – the experience on the wet street becomes so much more unpleasant than any experience in any warm restaurant could possibly be – so we enter a place we've already walked past at least three times. A *Bierkeller* of some kind. With the inevitable robust waitress in a red dress and white apron.

'*Noch ein Moment*,' I tell her. And to kill the time. '*Zwei Pils, bitte.*'

She doesn't smile. She's used to other kinds of interactions.

'We need a smartphone,' Ele says.

But I'm holding out against smartphones. I'm not even sure why. I'm the kind of person who holds out against a novelty until absolutely everyone else has it.

Once again we study the menu. *Kurpfälzer Wurstsalat mit Treberbrot. Schweinskopfsülze mit Remoulade und Bratkartoffeln.* The words are written in a Gothic typeface, which I suppose fits in with the traditional dress and the dark panelled walls and the wrought-iron lamps. Although we can read the Gothic letters they somehow make the menu even more forbidding. No loops are connecting. As soon as words become opaque and incomprehensible, you realise how absolutely crucial language is and how much of our perceptive experience is mediated and transformed and remembered and shared with the aid of language, so that any account of human consciousness is going to have to say something about language's role in it, even if almost nothing I have read so far has convinced me in this department.

Eleonora knows no German at all, and so is relying on my C at O level forty-five years ago. Unfortunately, there is no '*Schafkäse* moment' on this occasion. All I know is to steer clear of the *Wurst* and the *Schwein*. In the end – as we see our beers approaching – I suggest something for her that I believe is lamb, since she likes to eat meat when we're away from home, not being vegetarian herself, and likes to eat lamb when she's not in Italy, since the Italians only really eat lamb at Easter. For myself I go for *Bauernpfanne* (*Gemüse – Bratkartoffeln – Käse*), which seems to be meat-free. What can you do?

Raise your glasses to start with! The beer at least closes a healthy loop of recognition. You can drink German beer all over the world. We've been there. The experience is reassuringly renewed.

'The problem for me, with the enactivist position,' I'm telling Eleonora, 'is that although it's convincing about the force field that consciousness evolved in, us trying to meet our needs, looking for a restaurant, or even us finding each other, if you like' – she smiles at this – 'it doesn't even try to explain what it, consciousness, physically *is*. For example, it's not clear what Fuchs means when he says "the object is part of the conscious experience", since objects, in the enactivist view, are still considered in the traditional way, as things you can measure and study quite separately from yourself. In what way, then, are they part of the experience? I pick up the fork to eat my food. It "makes sense" in my hand, as it were, I feel it, it fits, and I know how to use it, but in what way is this particular fork now more part of my consciousness than everything around it, the room, the waitress and so on. Does it change when I touch it to take on a consciousness that includes my neurons in the famous closed circuit? Is some kind of electricity passing through it? How is this happening? Will the same thing happen when I just see it rather than use it? Isn't it perfectly possible to be using a fork while unconscious of it, or hardly conscious, since I'm concentrating on a conversation or a television programme, etc.? And in general why does so much perception occur in places where we have no interaction and could never ever have had any interaction? The moon, the stars. At the end of the day, if the object only exists insofar as it is part of a possible action, there does seem to be a wild excess of perception with regard to activity. A girl turns a boy's head. As they say. That makes sense in enactivist terms. She excites a few million languorous loops—'

'Pig,' Eleonora interrupts.

'—but why can he see the complicated pattern on her Armani handbag? Fuchs tried to explain, but I wasn't convinced. Then again, you can take the same approach, I mean of studying consciousness as a product of our evolution, and yet end up with a theory for consciousness being inside the head, or maybe not even really existing at all. Daniel Dennett does that in a book called *Kinds of Minds* where he starts with the most basic life impulse, the smallest organisms, and argues how an essential intentionality – inclination towards engagement – in all living things could evolve into what we call consciousness; but he doesn't think consciousness is shared between brain, body and object. The body just becomes an ever more sophisticated input–output device while the brain evolves to process the incoming information as consciousness.'

Eleonora has been listening carefully. 'Isn't there a problem with the word action, too?' she asks.

'In what sense?'

'Well, Fuchs is saying mind is created by action, isn't he, but don't you need a mind already, to initiate an action? I mean, when a machine does something, do you say it performs an action? No. Or when a flower closes at night or opens in the morning? No. It's simply automatic. So when you say something is an action, like looking for food, you mean there's intention, which means there's already mind. You can't start with action or intention then arrive at consciousness, since you need consciousness of some kind, however primitive, to act or intend.'

It's a good point. The truth is that like all Italians who went to school at a *liceo*, and particularly a *liceo classico*, Eleonora has studied philosophy and logic. Which I never did.

'It seems,' she's saying, 'from what you're telling me, that they talk a lot about things around consciousness: evolution, intention, the environment, behaviour, neurons and their billions of connections, without talking about the thing itself.'

'Right. And it's odd, isn't it? Most science isn't interested in the *correlates* of something, but the something itself, even when it's an incredibly elusive subatomic something. The Higgs Boson, for example. They're not satisfied with Higgs Boson correlates. They build a hugely expensive nuclear acceleration tunnel to find the thing itself, Higgs Boson. But that doesn't happen with consciousness. They describe a lot of fantastically intricate stuff going on in the brain, without making it clear how it links to consciousness, except to say it happens at the same time, or that the experience can't happen when that part of the brain is damaged.'

At which point our food arrives and though my dish is more or less what I expected, Eleonora's is quite different, more abundant and elaborate than we could have imagined. Not lamb but a mix of boiled meats in a broth of greens with a truly huge, triumphantly Teutonic quantity of mashed potato on a separate dish that I immediately realise will largely be switched to my plate. Eleonora just laughs. We raise our glasses again.

'To *Bierkellers*!'

'Maybe your neuroscientist lady tomorrow will be enlightening,' she says.

'Of the three she certainly has the largest research budget!'

We eat the food which is quite different from what we're used to and hardly attractive to look at, but not bad in the mouth. And after ordering a second beer, I tell Eleonora that amazingly Fuchs, having had trouble fielding the argument from dreams himself when I put it to him, didn't even bother returning the question when I explained the Spread Mind theory. 'Which is maybe a good job, since I'm anxious I won't be able to explain.'

'Try it on me,' she says.

'You're too sympathetic.'

'I'll play devil's advocate. I'll be vicious.'

'You couldn't be vicious if you tried.'

'We'll see. Try it.'

Every time I set out to explain the position of dreams in Riccardo's Spread Mind theory it's like negotiating a difficult rapid in my kayak. Obstacles appear out of nowhere and you're not sure whether to go round them or through them, or what you're likely to meet on the other side.

'The short version or the long?'

'The short first,' she says, 'then the long.'

We do have a lot of potatoes to get through.

'Short version. You never dream anything you haven't already experienced as direct perception, and when—'

'Stop. Not true. I often dream weird scenes of things and people that couldn't possibly have existed. Didn't you dream a headless dinosaur walking up a stream this morning? You've never seen a dinosaur.'

'Obviously there can be *mixes* of old experiences, lion's head Photoshopped on dog's body sort of thing. Pink elephants. Donkeys with angels' wings. Think of the way you see reflections in a shop window superimposed over the displays behind. It looks like one thing, but it's actually a convergence of all kinds of bits and pieces. Descartes pointed this out. That fantastical animals in dreams and pictures were always put together from parts of other animals.'

139

'I thought we didn't agree with Descartes.'

'We agree with him on this. You can't dream of a colour that doesn't exist, for example. Or dream a sound that's never been heard. Or a perfume no woman ever wore.'

'Hard to prove, isn't it? What if I say I dreamed of this colour no one has seen before? How could you prove I'm wrong?'

'I couldn't. But I appeal to the weight of collective evidence. People don't claim to dream of new colours, smells or sounds. Or hallucinate them. For example, a guy called Wilder Penfield managed to get people to have hallucinations by stimulating parts of their brain electrically during open brain surgery. He took this to indicate that experience is secreted by the brain and that the brain could produce experience without an external world. But on examination the hallucinations the patients reported were all clearly things they had experienced before. Their wife's voice. A room they knew, whatever. Mixed up as if in a dream.'

'Memories stored in the brain!'

'We'll get back to that. Let's stay with dreams, for the moment.'

'The short version.'

'The short version: you never dream anything that hasn't in some way formed part of a direct perception. All the same, these things you dream, however concocted and rearranged, are *not stored in the brain*. And they are not loops referring back to past experience and randomly activated in sleep, which would only really amount to a fancier kind of storage.'

'So where are they?'

I take another slug of beer. This is the point where the

Spread Mind theory becomes a real challenge, requires the kind of courage Galileo needed when he declared the earth was spinning while everyone else *knew* it was rock still.

'Have you ever thought,' I interrupt myself, 'that the inherently conservative nature of society, the fact that it's always trying to keep everyone in line and collectively deny anything that would upset its premises, so that it can remain what it is, means that anyone who puts forward a revolutionary idea has to be slightly mad, perhaps megalomaniac, in any event willing to stick their necks out. I mean, you can imagine someone realising the earth is not the centre of the universe, or that the planet is infinitely older than the Bible suggests, or that the human species is not made in God's image but evolved from an amoeba, and simply saying to themselves, hey, let's not go there, they'll only give me grief. It's not unlike the situation where you're reviewing a famous writer, say Salman Rushdie, or Elena Ferrante, and you realise their work is not very good, actually it's seriously bad, but also that if you come out and say that, you're just going to make trouble for yourself.'

'Dreams!' Ele laughs. 'Explain!'

'Sorry, it just occurred to me that Riccardo has to live in a sort of delirium of self-importance even to imagine he might be right and the textbooks wrong.'

'Get on with it!'

'There was a guy called Hermann von Helmholtz, for example, who said that anyone who comes up with a new concept "finds, as a rule, that it is much more difficult to see why other people do not understand him than it was to discover the new truth".'

'Enough!'

'OK. Deep breath. The short version. You never dream anything you haven't encountered in direct perception where, remember, in the Spread Mind view, the object *is* the experience. The experience is located at the object. Brain and perceptive faculties make it possible for the experience/ object to manifest itself. Simply, the world is what you see. That is conscious experience. And in dreams *this continues to be the case*. The dream experience is located at the object. Not in the head.'

Eleonora chews her boiled meat. More seriously than before, she says, 'That is quite mad. Or you explained it very badly.'

'It was the short version.'

'So give me the long version. I'm beginning to worry for your sanity.'

Of course, on the one hand I'm using this remembered reconstructed dialogue to try to inch towards the ideas underpinning the Spread Mind theory, bold ideas that may or may not make sense. We'll see. But at the same time to express my happiness with Eleonora, and to relish the part of my life I have spent with her. We love talking about stuff like this. There is a huge gap of years and experience between us, yet we mesh in a way the enactivists could only applaud. Each seems to offer the perfect environment for the other. For example, we both thoroughly enjoy whatever reading the other is doing. Not that we read the same things, we don't, but we talk about what we're reading. And discuss and disagree without arguing. It's as if, in a way, this relationship, these daily conversations where every subject is treated as urgent and interesting, had made me hungrier for new ideas. Because they feed our relationship. There is something physical about it all that has to do with appetite, and conation. It's the opposite of depression as Fuchs described it. A sort of enhanced health, a crackling fire that likes plenty of fuel. I find myself reading *in order to* discuss things with Eleonora, and of course Riccardo. I feel very lucky that this situation has developed. One needs a sort of mad energy, an insane optimism, to turn effort and attention to an ancient conundrum like consciousness.

On the other hand, excited barking does not indicate that man's best friend has found the right tree.

'My sanity?'

'I'll reserve judgement till I've heard the long version.'

'OK, but the long version is kind of long. Before I launch into it, why don't you tell me exactly what seems mad about the short version. So I can focus on that.'

Eleonora sighs. The *Bierkeller* is getting busy and we're having to raise our voices now. Perhaps her ear is troubling her. Not ideal conditions for philosophical debate.

'In your dream this morning there was a dinosaur. So in the Spread Mind version of the world, this morning while you were asleep a real dinosaur was the object of your experience. The experience wasn't in the sleepy head. It was the dinosaur.'

'Yes.'

She shakes her head.

'Case closed. Or rather, psychiatric case opened.'

'The dinosaur was headless.'

'Like the guy who invented the theory.'

'Of course it would have been a movie dinosaur, rather than a real one. I mean, we've all seen movie dinosaurs. They are part of our experience. We all know the kitchen scene in *Jurassic Park* with the velociraptor, or the bucolic bit where the big vegetarian dinosaurs chomp on the treetops.'

'Where is all this stuff coming from if it's not stored somehow? I mean, we're not seeing the movie now, are we? It's in our heads.'

'So that's your only objection?'

'It seems like a big one.'

It's fun discussing this with Ele, but what would it be like with Thomas Fuchs? Or tomorrow with Hannah Monyer? She wouldn't give me the time of day. Or a reviewer of this book? Perhaps after tomorrow's meeting with a real scientist I will return to Milan converted to the neuroscientific view, freed from the madness of Manzotti.

'Let's start with time. Textbooks tell us that we live in a linear progression of discrete instants of time, as Newton suggested. The world is there, entire unto itself, in separate instants, like atoms of time, one after another. But does that really match with our experience?'

'You're not going to tell me time goes backward, like in some experimental novels?'

She knows I hate those novels.

'Not at all, just we don't *experience* it in instants. It's more elastic. Just to understand a sentence, this one for example, to perceive it as a unit, you have to be capable of an experience extended in time. To understand that a clock has struck six you have to group its six chimes in a single unit.

Otherwise you might imagine that it had struck one six times. That was the basis of a long discussion between Descartes and a priest, by the way. And when you look at the world around you, everything is a different time-distance away from you. It may take only a millisecond for the light bouncing off my face to reach your eyes, but hills on the horizon will be a few milliseconds more, the moon is 1.3 seconds, the sun is over eight minutes away and the stars are years off. The constellations never really have the shape they appear to us to have, but only when we see them now at their different time distances away from us. That is, their shape is their shape to us now across the centuries. Which isn't an illusion, but just the result of the light from them taking vastly different times to reach us.'

'All you're saying is that what we see is already in the past, so presumably separate from us.'

'That's the logic of the textbook position. And it supports the idea that experience is in the head. Time is linear. So if what I experience is past, there must be a separate representation of it in the head, since what is past can't be identical with what is present.'

'A return to sanity.'

'Except, as we know, no one's identified representations in the head, and even in the head, different perceptive processes take different times to complete. Semir Zeki has written about that. Hearing, seeing, touching, tasting and smelling take slightly different times to happen. So that actually if you froze time you couldn't have experience of any kind, because it's made up of things constantly moving; it's more like a piece of music than a painting, it has sense through time, or rather it only exists in a process of constant change.'

'What's this got to do with dreams?'

'*Pazienza*! Let's imagine a world without clocks going tick tick tick, pretending that everything stops and starts again every instant. Your experience, day to day, hour to hour, is caused by, or identical with, stimuli different time-distances away from the activity in your brain and your organs of perception. Actually, the neuroscientists agree on that. Wherever the lapse is short enough to allow us to interact with the object – hit it, eat it, talk to it, escape it – we consider that *the present*, the here and now, even if it is a slightly spread now.'

'What do you mean, a spread now?'

'Good question. Consider killing flies. Something I used to do, but now don't.'

'You don't kill flies?'

'Now *you're* digressing. The problem killing a fly is that by the time your hand reaches it, it has moved. It is so quick that though it's near enough for you to think you're in the same time instant together, it's already somewhere else by the time you can hit it. To solve the problem, you wait till the fly settles on a horizontal surface, then clap your hands about two inches above it. Although it can move before you strike, it can't resist the upward flow of air caused by your hands coming together and, fatally, is drawn to a point where it coincides with your instant as your hands meet.'

'Yuck.'

'Right.'

'And that's a spread now?'

'I'm just saying that experience involves change, and change, even where it's the simplest interaction with the world, isn't instantaneous. Nothing is experienced in an instant, so nowness is always flexible.'

'And if we can't interact?'

'If we can't interact with an object, that's because it's far away or in the past, which to a certain extent is the same thing, distance becoming time when travelled by sound waves and light waves. But that doesn't mean we stop experiencing it. Not altogether. With distance things are clear enough. We experience the light of a star that set out towards us from millions of miles away many years ago. Or just a mountaintop fifty miles away. No problem. With our own past things are trickier. Suppose, eyes closed, thinking of you, I have some feeling about how nice it was having our afternoon nap together, hugging, kissing, etc. Well, it's not the same intense experience as when we were lying down, it's not even a semantic memory, I mean I don't deliberately recall the experience in words; it's more as if alongside and after the initial intense experience there was a sort of slow-release experience, as if some of the perceptual processes, instead of firing off immediately in order to interact, were delaying things as long as possible, so the experience will last.'

'Fantasy.'

'Think of when you go and see your family for a few days. They wind you up and you get furious. Then when you come back, for a couple of days it's as if that experience were still going on. You're still back in Taranto arguing with them even if you're in Milan. You're not the same person, but gradually you return to yourself.'

'Calumny.'

'I'll try to sum up. At any moment, your experience or consciousness is made up of things all kinds of time-distances away. Obviously, wide-awake attention privileges the stuff coming at you as you walk down the street, just a milli-second away – we all need to cross the road – but in your experience as you walk there's something you have to do at the post office and something your mother said on the phone this morning. That conversation is still unfolding. No experience is instantaneous, closed in the split-second point of the present. All perception is spread in time. You could think of it as a guitar string that's been plucked and is still vibrating. And when the thing perceived is far away – like a star – there's no question of interaction. Perhaps it's already extinct. You couldn't even plan a mission to it. So where's the difference from memories? Aside from the quality of the perception?'

'Come on, we know perfectly well the difference between memory and perception. We don't get them mixed up.'

'We do in dreams.'

'But then we wake up.'

'And say it was all in our heads.'

We're looking at empty plates now, time has passed as we spoke, or mainly I spoke, and I'm not sure how much has been achieved. An idea has been fielded, no doubt ill-defined; it has taken a while to bring it into being and it is already slipping into memory, unless I keep working on its construction. Every discussion is a process of becoming. This is a sandcastle that will need a lot of shoring up. Against the tide of time and other people's objections.

'I want a sweet,' I declare. 'I *demand* a strudel.' The only German sweet I know.

Eleonora agrees to share.

But amazingly this *Bierkeller* has no strudel. So much for stereotypes. Under pressure from the waitress, I settle for a *Kaiserschmarren mit Apfelmus*. I understand the word *Kaiser* and I understand the word *Apfel*, but how they're going to sit together on a plate I have no idea.

'Like a dream,' Eleonora suggests.

'Two spoons,' I ask the waitress. When she looks blank, I recall the word *Löffel*. '*Zwei Löffel*.' I even remember it has an umlaut. But after the waitress has departed, always with the same grim look on her face, I wonder whether I should actually have said *Löffeln*. Don't German plurals sometimes end in 'n'? The word spoon, given the special context of a German waitress in traditional dress with a blank expression on her face, has called up the word *Löffel*. But let's not get distracted by that kind of talk again. Before the sandcastle dissolves and the *Kaiserschmarren* materialises, let's nail dreams.

'There's an amazing passage in Oliver Sacks's *Mind's Eye* where he talks about a weird experience he had, presumably due to a tumour in his right retina. He found he was still seeing things for a while, *even after he shut his eyes*. He talks about a washbasin. He washes his face and closes his eyes while he's drying himself, but the washbasin is still there in front of him. Same in Times Square. Even if he closes his eyes, the neon keeps shining behind his lids. And again with people walking down the street. They keep being there even when he closes his eyes. This went on for a few weeks.

Whenever he shut his eyes the normal perceptive experience would continue, for about fifteen seconds, he says.'

'Strange.'

'Good for action replays. Close your eyes and see the goal again.'

'And so?'

'Being an internalist, endorsing the orthodox view of consciousness, Sacks took this to be proof that it is the brain that actually produces experience.'

'Makes sense.'

'Well, it couldn't go on producing it *after* fifteen seconds, nor could it produce it without seeing the washbasin or Times Square in the first place. And where it was a question of things that moved, when he shut his eyes, the experience would go on, but with the person running on the spot, so to speak. When he opened his eyes again, the world had changed and the person had moved on. His brain wasn't producing the *continuation* of the phenomenon, just holding it there for longer than it normally would.'

'So?'

'So, it would seem that when the body is operating normally, there are mechanisms that shut off what is no longer in front of you, presumably because if it stayed there it would get in the way of the next perception, it would stop you interacting with the world. All the same, this weird experience suggests that physically, if this suppression mechanism should fail for some reason, as in Sacks's case, there is nothing to stop perception from being released more slowly and so going on, after your eyes are closed. Perception delayed or retarded or extended. Spread through time. In the end, at a less dramatic level, our experience is always full of things going on long after the event. What about when you can't get a piece of music or a song out of your mind? Maybe something you don't even like. Remember in the film *Touching the Void*; when the guy is completely wrecked and crawling down from the mountain top he keeps hearing an old song, "Brown Girl in the Ring" by Boney M. He hates the song and he hates Boney M., but all the same the song keeps going on so that he thinks he's going to die listening to Boney M.'

'And so?'

'In a normal state of wakefulness that stuff is blocked out, or in the background. But in the film, which is a documentary remember, our hero's in a delirious state, has a broken leg, has been exposed to freezing temperatures for three days, not to mention traumatic experiences of every kind. So, as in the case of Sacks, the normal inhibitory processes aren't functioning and the music goes on and on. His perceptive faculties are picking up a reality that is still part of his present like a distant star is part of our present, our spread now. "Brown Girl in the Ring". Not memory, but real contact with the song.'

'I'm sorry, but how can it be? The song isn't there. No one is playing it in the Peruvian mountains. It can only be in his head.'

'Let me offer this analogy. Think of water flowing down a rocky river. The water plunges down, but behind every rock there's an eddy; that is, some water gets trapped behind the pull of the rock and turns round and round and round there. If a piece of flotsam and jetsam falls into the eddy it may be trapped there for months, even years. Except, when the river level falls, the tension around the rock eases and the water of the eddy remixes with the main current. The twigs and Coke cans and fishing floats can head on downstream again.'

'So?'

'So. Imagine the photons entering the eye and setting off the chain reaction down the optic nerve and into the neurons. There are actually quite a few connections to be made before you complete the circuit that makes the experience of the object those photons are coming from possible. The connections take a few milliseconds, added to the milliseconds it took the photon to get to the eye in the first place. So the object that is your experience is always in the past relative to the body and the neural activity that makes it possible.'

'You've already said all this.'

'But now imagine that beside those direct connections that complete the circuit at once, there are also any number of eddies, clusters of neurons where the signal set off by the photons continues to circulate, without re-entering the main stream. Any number of signals going round and round. There are trillions upon trillions of synapses in the brain, remember. Until, with sleep, the stream of immediacy, daily consciousness, lapses, and then from all these eddies, other signals are potentially free to emerge, in the world of dreams.'

'Well . . . it sounds rather like stored memory to me. It's in the head.'

'The neural activity is in the head. But the object and experience is where it always was. Just that the process allowing the object to be present has been drastically slowed and interrupted, fragmented, mixed up with other processes involved with other objects. They're simply more in the past than they are in waking consciousness.'

'And that's dreams?'

'That's the spread mind, in time. The essential idea is that there is never anything between you and the object of your experience, past or present, no images, no representations. You are your experience and your experience is the world you move in. Dreaming and hallucinating are just delayed experience. The world you move in is you. As life unfolds, over the years, that experience accumulates so that in the right circumstances – a lapse of immediate perception – all kinds of distant perceptions reappear in delayed reconnection with the object of perception. Only when you die is it all definitely gone. If you think about it, even the verb tenses, in English and Italian, recognise this. In English, for things that happened in the past, but in a time period that still continues, that you're still in touch with, you use the present perfect tense. "I have done this a hundred times in my life." But "Napoleon did it a thousand times in his." Whatever the "it" is. Where you use the present perfect you suggest there is this ongoing present, the past in the present, elastic time. And in Italian you actually use the present tense. *Siamo insieme da tre anni*. We *are* together since three years. You deny that there is any past at all in this.'

'Now you're trying to keep me sweet.'

'Three and a half years to be precise.'

'Stay on the subject.'

'I'm just saying, while you're still alive, your past is also in a real sense alive. We've agreed haven't we that our perceptive systems are never in error over what they perceive. Only the way we interpret perception can be wrong. The mirage in the desert is a real perception of a shimmering, rippling surface, but misunderstood. You can also misunderstand an experience because something is far away, mistake a horse for a cow, for example. Same with our present experience of the past. It's a real experience challenged by all kinds of distance and neural complications. And then by the sheer amount of experience that has piled up over the years. What some people call karma. The effect of a distant or long-past event on us is going to be smaller, more blurry, than an immediate and similar event. But it's still real. Until of course you die and then all the accumulated experience, all the time and the karma, goes with you.'

This thought induces a moment of solemnity. My mother died last year and I have been very aware of a whole mood and feeling, a wealth of mannerisms, recipes, gestures, sayings and smells that went with her. She and her experience are no more. It's disturbing. But now the *Kaiserschmarren* arrives to cheer us up. It's a kind of scrambled pancake, served with puréed apple and great dollops of cream. No sooner is it on the table than I have the distinct impression I have eaten this before. Perhaps when I worked on my farm in Switzerland. God knows where the *Kaiser* comes in, though.

'Perhaps that's why I love kayaking so much,' I tell Eleonora. 'Because when you are going down a rapid, you really are all in the present, the time spread comes down to what is absolutely proximate.'

'Unless you get trapped in an eddy!'

'True. Some eddies can be really hard to get out of. Even worse are "holes", where the water plunges down over the top of the rock and churns white beneath it. There are rivers where you have to be manically focused just to avoid the holes and stay upright.'

'But then when you get out of the stream the experience goes on in the holes and eddies of your mind.'

It certainly does. I often dream I'm kayaking.

And as it turns out, this dinner too is going to go on for a considerable part of the night. My digestion is no match for the *Kartoffeln* and the *Kaiser*.

The Mind's Gate

Are we to trust our experience as we know it and live it? Trust that it is a real something. That it matters. It *is* matter. Or do we put all our faith in measurements and numbers, in what can be seen through a microscope, demonstrated in an accelerator tunnel, objectivised, proven? As they claim. This is the question. The Higgs Boson exists, but colours don't. This is the present orthodoxy. There are black holes, but no smells. We are in the Platonic cave and need instruments of every kind to look at the higher reality outside, though, even then, what we actually experience are only readings on instruments. We are trapped on one side of a Cartesian duality, wondering what's on the other, constructing a hypothetical 'reality' in figures, predictions, 'ideas'. Or as the inventor and fantasist Elon Musk claims in an updated version of the same story, we live in a simulated reality, not a 'base reality', in someone else's computer game. We can't know what the world outside our world is like.

A great deal is at stake here. Not just epistemologically, but socially, psychologically. The moment I accept that my experience is insubstantial, simulated, removed from what

Elon Musk calls 'base reality', I open the way for someone to tell me what base reality is. You think the grass is green, but it is not. You think you are cold, but in fact you're warm. You think you decide, but in fact you are manipulated. You think you are in pain, but actually this is not the case. You think you are happy, but happiness does not exist. And, of course, if my suffering is false, so is yours. So how can it matter what pain I inflict on you, if pain isn't real? And so on.

I am repeating myself, I know, but it was very much in these terms that, on waking three or four times through this second night in our three-star hotel in Heidelberg, the previous evening's conversation continued to unfold and ramify alongside the rumblings of heavy food in my gut. Repeating oneself is what thought is all about. Yesterday was encroaching on today. Today is a stretching of yesterday perhaps, the present a gathering of the past. There had been dreams too. Something about a flat I owned somewhere but that I never visited and for which I had lost the keys. Even the address. Where was this place? Who knows?

Around 4.30 I got out of bed to pee and to meditate. This is something I do most nights. There comes a point when I will wake and find it difficult to fall asleep again. I feel uncomfortable, restless. An hour's meditation sets me right, after which I return to bed and sleep like a log. People are disconcerted if I tell them this. They assume some kind of admirable renunciation on my part, a dogged asceticism. Or unhealthy fanaticism. In any event, something that sets me apart. Above all they ask me *what time* I get up. They want to know the hour, hear a *number*. They ask me if I set an alarm. My extremity is to be measured by how early I climb out of bed: six begins to be austere, five is heroic, four is monastic, three psychotic. And so on. We are so trussed up in the clockwork we have created that we suppose a particular reality is to be attributed to these digits. The hours of the clock face.

I reply that I do not set an alarm for my meditation hour. The very word alarm alarms me (it's Ele who sets the morning alarm). What matters is my perception, at whatever time I happen to wake, that given the various discomforts, or excitements, I feel I am not going to go back to sleep. The digital measurement of supposedly discrete instants of time according to a system invented a few hundred years ago and only universally imposed when communication over distance made it necessary to have some shared grid to place over the flow of our experience, simply does not come into it. I wake a first time, go to the bathroom and know I will fall right back to sleep. I wake a second time, go to the bathroom and know I will *not* fall asleep however long I lie and fidget and allow the previous evening's drink, talk and food to make their presence felt. This knowledge is real in a way the hour 4.15 is not.

So, when at home in Milan, I get up, boil some water, make a camomile tea, cover the cup carefully to preserve the heat, meditate for half an hour, drink the camomile tea, meditate another half an hour, more or less, and return to bed.

But what do I mean meditate, and why bring it up at this stage of our discussion of consciousness?

I was drawn, kicking and screaming, to meditation, some

ten years ago, in response to chronic abdominal pains. It's a story I tried to tell in the book *Teach Us to Sit Still*. Somebody had suggested that learning how to breathe in a different way, and thus deeply relax the body, might be a solution to my problem, or at least something that helped. The best way to learn, he said, was to take a course of vipassana meditation. I put the word vipassana in Google and Wikipedia responded thus: 'Vipassanā (Pāli) or vipaśyanā (Sanskrit: विपश्यन; Chinese: 觀 *guān*; Standard Tibetan: ལྷག་མཐོང་, *lhaktong*; Wyl. *lhag mthong*) in the Buddhist tradition means *insight into the true nature of reality.*'

The true nature of reality! What could be more pertinent to our discussion? According to an ancient tradition, then, when I get up in the middle of the night and practise this technique – for technique is what it is – I am gaining knowledge of the true nature of reality. A knowledge, it is implied, that cannot be gained *intellectually* from books, a knowledge that is not *information*, but on the contrary a setting aside of information. Not the knowledge of Galileo, then, and Bacon and Descartes. Or of Christof Koch.

But if vipassana knowledge is not information, can anything useful be *said* about it, in *words*? Can it become part of our discussion? Can it be expressed? For although we may respect a knowledge, wisdom even, that is incommunicable it would hardly make sense to write books about it. We want to be able to *say* what conscious experience is, not just talk around it. Perhaps the way to proceed, then, is to ask: how does this experience of meditation sit, to risk a pun, with the various theories of consciousness we have looked at?

At home I have a mat by the sofa in the sitting room, with

a hard low cushion – a *zafu* – and a thin blanket that I can put over my shoulders. Why not just sit on a chair, you ask? Because the position is important. There is nothing exotic or magical about this. Simply, the cross-legged position with the cushion to tilt the pelvis forward allows the back to sit straight and relaxed without being in contact with anything. Maintaining the position is not too hard (when you're used to it!), but requires constant wakefulness. You can't do it if you're distracted. So this position is a guarantee of wakefulness. And wakefulness is what vipassana is all about.

But wakefulness with regard to what? And what is this wakefulness *for*?

In the hotel I don't have my regular mat, or my *zafu*. I don't travel with them. In general I try not to become dependent on particular props, and I frequently vary things depending on how I feel. Perhaps a couple of yoga exercises instead of the second half-hour. Or no second half-hour at all. Or an hour and a half. As it comes. I don't want this to be a straitjacket, an obsession or a routine that I'm afraid of straying from.

So I put a quilt on the floor between the bed and the bathroom, roll some towels into a cushion and cover my shoulders with my overcoat against the cold. I don't exactly cross my legs but point the knees outward with calves and shins lying one alongside the other, something those in the know call the Burmese position. I close my eyes, draw two or three deep breaths to straighten my back and drop my shoulders, place my hands palm up on my thighs, and I'm ready to go. Or rather not go. Where could one possibly *go* in this position?

The supporter of an internalist view of consciousness would

ask, what is the relationship between the neural activity in my brain and the things I am feeling or hearing, my perceptual experience? Which neurons are busy producing the sensation of breath going in and out of my nose, which those of ankles pressed on the floor, which the initially rather painful stretching of the thigh muscles, which the tension in my jaw, which the thoughts that come to me in the form of words, which the sounds (of a car passing in the street, someone flushing a loo in another room)? And so on. The experience, for the internalist, *is* neural activity, or is produced by it, secreted by it, and in this regard meditation is no different from any other activity.

Internalists, that is, say little about the *nature* or *quality* of the conscious experience and restrict their talk to whatever they can show is going on in the brain at the same time. It is true, in that regard, that neuroscientists have excitedly demonstrated that the expert meditator generates gamma waves in the brain – or to be more precise, that when the expert meditator meditates gamma waves occur – a regular oscillation of neural firings at a frequency around 40 Hz, something that seems to be associated with an experience of well-being. But, as Sabina Pauen told us, there is no real agreement on the function of gamma waves, and no sense of why or how they might correlate with well-being; do they cause it or are they caused by it? If, of course, during meditation, I should have some kind of visual experience, despite my closed eyes – seeing a face, for example – internalists would remark that this is proof that experience in general is generated internally in the brain and does not require an outside world to manifest itself.

I say, seeing a face, because there was a period when, after perhaps half an hour or forty minutes of meditation, I would find myself seeing very distinct faces, usually of people I did not know. This happened in particular when I would go to a retreat where, typically, you sit still, eyes closed, in silence, for many hours of many days. The faces would 'appear' in an upper area of vision, as if looking down on me, from a painting in a church for example, before dissolving after a few seconds. I was so curious about this phenomenon that I went to see a vipassana teacher and asked him about it. He was unimpressed. 'Vipassana has nothing to do with seeing visions,' he said. 'In vipassana we are wakeful and focused, and in that state you do not see visions. You see visions when you lose concentration and the mind is sleepy or unfocused in some way.' My visions, in short, were an indication of my ineptitude as a meditator.

The teacher's explanation would fit in with the Spread Mind idea that so-called delayed perceptions – dreams, hallucinations – occur when wakeful, interactive perception lapses. Sleepy, or just musing in some safe protected situation, or under the influence of a drug perhaps, one sees and hears things 'that are not there'. One sees the past. Or rather, seeing these faces alerts one to the truth that the past is part of the present. That is, *my* past is part of my present, to the extent that my perceptive faculties are capable of allowing past events to continue to have an effect on them. Normally the mind privileges the world it is obliged to interact with, but this is not one of those normal moments, and faces appear. 'Images!' Semir Zeki, would say, secreted by your neurons. Dormant loops activated at random, the enactivists would say, old experiences waiting to be reconfirmed.

Of the three theories we've looked at, enactivism would seem to be the one with the least to say about this supremely inactive activity of meditation. I assume my position on my cushion, knees pointed outward, ankles side by side on the mat. Eyes closed, I run as it were a check on the various parts of the body to see that the gravitational pull is going straight down my spine to my cushion. More or less. Then I move my attention to my upper lip and simply wait until my body breathes and I feel the breath passing over the lip. I don't breathe *on purpose* or in any particular way. And I don't deliberately hold my breath. I simply disengage from any purposeful activity, except observation. In this way, all at once, reliable as daybreak, a breath from the belly kicks in. It's a big breath and is immediately followed by a certain relaxation of various muscles. Interestingly, it's difficult to maintain concentration across the arrival of these first deeper, reflex breaths; they are moments when thoughts from outside will stray in. So this relaxation would seem to go together with a disengagement from the world of interaction, a disengagement that allows the past, or delayed perception, or random neural generation of perception, or reactivated loops – take your pick – to click in. If you let them.

Perhaps what an enactivist would have to say is that the meditator is doing something unnatural, in that he or she is trying to disengage the conation, the essential life drive, that forms the day-to-day reality of any living creature. The meditator seeks to arrive at a state that is at once wakeful, all attention, but without appetence, without any reaching out to the world, that 'intentionality' that Daniel Dennett loves to track down. This is why the hands are placed palm up on the thighs rather than palm down. If you place your hands palm down, your instinct is to hold or grip the thigh, and if you do that then hands and arms and shoulders are activated. They become props for the back. Palms up, the hands are, as it were, disabled, disconnected. In this regard you could see meditation as a challenge to enactivism, since for the enactivist there is no experience without conation, and the meditator certainly has experiences, even if only in the area that Fuchs described as *Befinden*, the background awareness of your own physicality. The enactivist would have to say that the meditator fails to achieve his goal of entirely eliminating all drive and that what he experiences is a minimal active engagement with reality, the feel of the floor, of the compressed knee joints, of air on the face, etc., all offering possibilities of action which he abstains from, conation, as it were, perversely opposing conation.

The first minutes of meditation are not easy. Often, after sitting for some time, I will suddenly become aware that I haven't started 'meditating' yet at all. My mind has been entirely taken up with thoughts. They might even be thoughts like these, thoughts about consciousness and meditation. Certainly in the early hours of this morning – I mean today, the day of writing, not Heidelberg – when I started to meditate, my mind was crowded with things that I might want to put in this chapter. That is, I was aware in the early hours that later on I would be beginning this section of the book, and now, at about 9.30 a.m., I am aware that what I am doing, writing this down, is a continuation of thoughts that made my earlier meditation difficult. I had ideas then that I am elaborating now. This is typical. And the enactivist could smile now and say that this was conation reaching out to the action of writing. The internalist could say that the neurons were secreting their stuff. In this case thought, words. Though of course no one has ever found any words in anyone's head.

Even more often, though, the thoughts in these early moments of my meditation are jumbled and disordered. Perhaps there will be two, three, or even four areas of thought, often anxious thought, swimming in and out of focus. Then a moment of self-reflexive attention. Damn, I'm thinking about the Scottish referendum or about Wayne Rooney, instead of meditating. Or I'm listening to a sound in the street. Or wondering what a certain humming noise might be. And I attach words to the noise. Is it the fridge? Does this hotel room have an air conditioner? Then I realise I'm thinking these thoughts and turn my attention back to my upper lip, back to the breath crossing my upper lip. I focus on that. But now, damn again, I am thinking about my tax situation, or a colleague at the university who gets on my nerves. What is striking is how *continuously discontinuous* consciousness is. I didn't *decide* to start thinking about my tax situation, or my irritating colleague. But suddenly I realise I am thinking about him *and have been for some time*. It's hard to have a steady awareness, even of one's own thoughts, in the absence of an urgent object of concentration. It's hard to be constantly aware of what you are aware of. The breath on my lip is there, it's a sensation, but tenuous, it is not much to hang on to. It varies, but doesn't vary much. Not enough to keep me *engaged*. If one wants to concentrate, to experience a continuous state of focused awareness, it is easier if one has something intense, something constantly and urgently changing, to concentrate on, like an exciting river rapid to paddle, or a good book to read, or a film to watch. Even a simple task to be getting on with will do it; preparing an aubergine curry perhaps. But meditation encourages awareness of all that we are not normally aware of. To gain our insight into the true nature of reality, we observe ourselves in the present, now – with great attention and without any interference from without or from the past. We must observe our own attention in a way, our own struggle to observe. It's hard.

But why? Why does it have to be this way? Why is it so difficult not to fall back into old thoughts?

Once discussing this with Riccardo he suggested the analogy of a river that deepens over the years as it cuts its way through the landscape. We might call this the self, becoming more and more defined as, with each new year, it sinks deeper and deeper into the same path it always takes across the landscape, a path that has become one of compulsion and inevitability. But flatten the landscape and suddenly the water can go anywhere, take any path. Meditation is that flattening. Certainly, when briefly successful, it does seem that the self is quietened and the landscape new.

There are techniques for achieving this, the techniques they teach you at retreats. The worst thing you can do is to be irritated by your own inevitable failures, by the discovery that your mind has drifted back to politics, or work, or worries about money. Even the word 'damn' is a mistake. Irritation causes more thinking, introduces a negative judgement on yourself – I'm not able to focus my attention, I'm hopeless at this, etc., etc. Such self-criticism is a form of what the Buddhists call attachment, the mind's reflex habit of engaging with the world and latching on to it. There are clear analogies between this idea of attachment, the enactivist concept of conation and Daniel Dennett's 'intentional stance', in short our driven reaching out to and connecting with the world. Enactivism considers conation an indication of health; Buddhism considers attachment the beginning of *dukkha*, that is to say of pleasure and pain, but ultimately suffering. I don't think these positions are mutually exclusive. Both describe the deepening river of the driven self, with all its pains and pleasures. 'There is always an element of suffering in pleasure,' Montaigne remarked.

Sitting on my rolled-up towels, my legs a little stiff, my stomach rumbling, my overcoat not entirely protecting my knees from the cold, I'm very aware of *not* being at home, aware too that later on this morning I have to meet a highly respected neuroscientist and engage in a debate involving facts and concepts that are still extremely difficult for me to grasp. The strategies available for overcoming these anxious intrusions are a quietly determined focus on the breathing and, alternatively, a body scan. This is perhaps the best-known meditation technique. You move your attention away from the breath on your lip and through your body, through every part of it, every inch of your skin, if you can, from the top of your head to the soles of your feet. You can do this at speed: one breath your head, one breath your shoulders, one breath your arms, one breath your hands, and so on. The rapid movement can help you to maintain your level of attention. Or you can do it slowly: a long focus on the tiniest stirring of air on the scalp, the faintest humming of sensation in the temples, the tenuous awareness of the eyes in their sockets, the rather stronger tension in the palate and throat.

What kind of perception is this? It's not seeing, because I don't see or try to 'visualise', whatever that word means, anything. It's not hearing. Of course there are sounds, the occasional sigh, a faint noise of traffic, but this is not where my attention is focused. It's not smelling. Entirely attuned to whatever the smell of the room is, I smell nothing. Smell almost always comes as an awareness of change, it rarely has continuity. When there is no change in smell one has no awareness of it. It is definitely not taste, since there is no taste in my mouth.

What about touch? To an extent I am experiencing standard perceptions of the kind we call touch. For example, I may initially feel for the contact of clothes on skin, or the cushion on my butt, or the quilt under my ankles, to help me focus my attention. But there are also scores of other sensations that don't involve anything outside my body, or even the surface of the skin. Sensations of tension, of heat, of prickling, of glowing. They are different in different parts of the body and as I pay attention they intensify. Or rather, I may concentrate on my hands, and apparently there is no feeling there. Then, one by one, as I focus on them, the fingers light up, as it were, and glow. They feel warm. They begin to pulse a little, to throb. I may even feel a pulse wave that seems to move from one hand to the other without passing through the body in between. Bringing wakefulness to bear on a part of the body for a period of time alters our perception of it, or perhaps causes the body part itself to change. In any event, it all begins to feel more mobile and fluid than you would have thought.

What reality do these sensations have? What insight into consciousness do they offer? Is this just, as it were, background neural firing, the grey matter ticking over? Or, in the Spread Mind view, is my body, the object of perception, identical with my experience, so that I am now knowing my body more intensely than I ordinarily would, the way I would know a stone more intensely if I spend an hour staring at it? Certainly, one is very aware of the intimacy of body and mind in meditation. Focus on your shoulders for a while and you realise they are tense, there is a life in them that is indeed a mix of pleasure and pain. Maintain that quiet focus and they might just relax. If you're lucky. Same with the lips, same with the hands. Even the thighs. If you are really patient, bringing a quiet wakefulness to the tension in your thighs, truly accepting whatever discomfort there is, not trying to change it in any way, then, curiously, they may, just may, finally relax. 'Finally' meaning ten minutes, thirty minutes. When you are in form, you can focus quietly on an area of pain, the back, the feet, just observing, not seeking in any way to intervene, remaining absolutely still, and the pain will eventually subside, or simply become acceptable, an entirely bearable part of a general flow of sensation.

This is a wonderful feeling. Perhaps it correlates with gamma waves. Perhaps it is an approach to 'the true nature of reality', as Buddhists claim. In any event, you mustn't grow consciously attached to it, or it is lost. You mustn't *attach words to it.* It is words, even more than pain, or itchy impatience that will distract you and disperse the meditation experience. You hear a motorbike accelerating noisily in the street. And you say the word motorbike. Or rather the word motorbike *says itself.* Then perhaps there is some fleeting thought about the absolute necessity for strict regulation of motorbike sound emissions. Framed in words of course. Then a reflection that such regulation very probably exists already, but is not enforced. Then you are wondering about the passivity of society in the face of louts and idiots. Is this an indication of tacit support for motorbike manufacturers, the hypocrisy of politicians who pretend to protect the public but are in fact hand in glove with their industrial cronies? Then you are criticising yourself for getting on your high horse. You are such a crusty old reactionary. And so on. All words. Until again, abruptly you become aware that you have been distracted. You must get back to your body scan. And when you do, you find everything hardened again. The flow has gone. One of the reasons for assuming the cross-legged sitting position is that it will become painful if you allow your mind to be distracted and your body to harden. So pains in your legs warn you that you have lost concentration.

Words are your enemy, then, while meditating. They do not bring you closer to 'the true nature of reality'. But in general, language profoundly alters the nature of perception. We attribute words to things as they happen, fishing those things out of the flow of immediate experience with the hook of a noun or name or verb and then setting them off on a train of thought suggested by verbal association; we remove them from one continuum and place them in another. Perhaps it is not entirely unlike the difference between a face seen and a face photographed. Reality is halted, in a word or a snapshot. It is possessed, then sent to a friend on WhatsApp, where it exists in another environment, subtracted from its natural spatial and temporal surroundings to become part of a discussion, or an album. I see a tree and instead of dwelling on its solid trunk and rustling foliage, the word tree comes to mind. As soon as I have the word I can make a statement. What a nice tree! And at once the phrase separates me from the experience. I don't savour the tree. I start thinking about trees. Which is to say the words set off where they will. Very likely a horse chestnut tree, I tell myself; perhaps there will be squirrels in its branches. In so doing I am bringing old experience and common knowledge to this immediate sensation. There were squirrels in the horse chestnuts in the vicarage garden of my youth. More than anything else, it is our verbalising of experience, and our translation of sensation into reflection and prediction that gives us a sense that the experience is internal to our brains. Where is this verbalising going on, if not in my head?

But if words alter immediate perception, they change our perception of the past even more dramatically. No sooner have we become aware of some old experience than words are attaching themselves to it like bees to honey. 'Once a thing is conceived in the mind, the words to express it soon present themselves,' wrote Horace. 'The things themselves force the words to express them,' observed Cicero. Complacently, we call the words *thoughts* and assume that we are in control and that there is purposefulness in our 'thinking'. But to a great extent thoughts 'think themselves', as Nietzsche said; they are compounds of past experience and words, hurrying along on the metalled ways that language, with its standard syntax and rhetoric, offers. So memory, deferred perception, is verbalised and, with verbalisation, abstracted from the initial sensation, shunted elsewhere. Naturally we can bring our conation, our instinctive drive, whatever we want to call it, to bear on these words and start taking them *deliberately* in this or that direction, just as we can choose to walk this or that way in the street. Language, we might say, opens a different map for us, the map of a thoughtscape which is very far from 'the true nature of reality', though inevitably it is part of it.

Perhaps the idea then – typical of many approaches to consciousness – that we, or simply the body, automatically privilege the immediate world around us, the world we can engage with *now*, is not altogether true. Or no longer true for the human being. Perhaps language, surely a late development in the evolutionary process, has given such intensity and presence to our experience of the past and such pleasure in our ability to manipulate it and to manipulate perception in general, in what we call thought, such a complacency in verbal agency and the so-called sense of self that accompanies it, that very often we are in fact *unaware* of things right here and now that we would do well to be aware of. So, walking down the street thinking of consciousness, I am unconscious of the uneven paving I am about to trip on.

Worrying about my imminent visit to the dentist, I do not pay attention to the traffic lights. Wasn't this how King Alfred burned the famous cakes? He was thinking of something else. He didn't notice that burning smell until too late.

The fact is we love this impression of being *inside* ourselves, abstracted, separate from the immediate physical world, that words give us. We reinforce the feeling by engaging constantly with objects and pastimes that intensify this supposedly 'inner life'. Everything becomes 'food for thought'. And where might this mental food be, if not in the head? Every time I ask my friend Riccardo about language, he says he is concentrating all his efforts on immediate direct perception and will come to language later. Interestingly, he claims that he verbalises experience very little, and suggests that it is his habit of seeing the world without compulsively attaching words to it that has allowed him to overcome the normal internalistic approach to consciousness. Internalism makes no sense to him because he lives in a world of external objects which are his consciousness. So he claims.

Yet it seems to me that this problem of words must be addressed, at least in some provisional way, since, for most of us, certainly for me, verbalisation is constant and it's hard to think of consciousness without admitting the presence of language. Aside from meditation, there are only rare moments of intense physical engagement, or complete distraction, when I am not verbalising, and nothing gives me an impression of internal selfhood more than verbal thought. It's true that if one looks back at the Cartesian 'I think therefore I am', one is astonished at the credit that was given to it. I can doubt everything, Descartes says, sight, sound, touch, smell, taste, I could be mistaken in all of them, but I cannot doubt that I am thinking. A child could point out that, however mistaken you might be about your five senses, nevertheless it requires someone to be mistaken. Even an illusion requires an entity that can misperceive. 'He might just as well have said, It's raining, therefore I am,' Wittgenstein once remarked, for all the difference it would have made, since to observe, rightly or wrongly, that it's raining, you need to exist. In any event, there is always something wearisomely formal and provocatively 'clever' about extreme scepticism – how can I know the world exists? – since whatever the answer, you are anyway obliged to get on with life as if the world was as you perceived it. In fact, the only conclusion one can come to is that 'I think therefore I am' took hold because it was so *attractive*. It was what people wanted and still want to hear. They mistook Descartes' claim for the notion that one is identical with one's thinking, something he didn't actually say, identical with the constant stream of words over which they believe they have mastery, not with their bodies, or with the world as it is perceived, over which one has so little mastery.

Let me give a banal example of how, at least in my case, language reinforces a general complacency about self and identity. A couple of days before writing this – I am writing, as I said, *about* September 2015 in Heidelberg, but *the time of writing* is July 2016 in Milan – I went to the swimming pool for the first time in almost a year. It's a big open-air pool and it was good to be back in the water, though after such a long break my movements felt awkward. At a certain point, aware of an unhappy rigidity, a sort of pulling myself – my head particularly – away from the water, I remembered, or sensed, that to swim properly I must let go, let my body be borne up by the water. Immediately, I said, Give yourself to it, Tim, let your head go, let your chest go. You must give yourself to the water. And this performed internal verbalisation of a physical awareness leading to or simply simultaneous with an adjustment of my behaviour in the water, gave me a sense of selfhood, of being a person. It created a kind of drama around an ordinary recreational activity, the drama of a sixty-one-year-old man going back to swimming after a year's break and feeling for the proper motions. Thinking, we congratulate ourselves we exist. We love Descartes for having centred our epistemology in thinking.

So do the internalists win? The insistent occurrence of words in my mind while I sit eyes closed trying to meditate in my Heidelberg hotel bedroom, confirms that consciousness is generated in the brain, the self is an internal thing. And that is that. Case closed?

Hume tells us that all experience is in one way or another perception. We have immediate experience of the world around us, which he calls 'impressions'. And we have later echoes and returns of those impressions in attenuated form, and these he calls 'ideas'. ''Tis impossible', Hume remarks, 'for us so much as to conceive or form an idea of any thing specifically different from ideas and impressions.' These attenuated ideas may manifest themselves in the form of language; nevertheless every experience remains connected, in however complex a fashion, to initial sense experiences.

That is, although Hume is an internalist, he shares Manzotti's view that there is no thought without initial perception of something external, and also that it is impossible to conceive of any object except as we perceive it.

We have already looked at the idea that memories might be thought of as a form of deferred or delayed experience, the perceptive system still intercepting the event, or still being susceptible to its causal powers even after a lapse of time. Could we then suggest that there is an initial sense experience of words as sound or signs – when we are first taught, or first hear or see them – and then their frequent return and performance in speaking, writing and thinking, their constantly meshing with fresh perception and delayed perception in the same mixing of diverse experiences that we get in dreams, every word we use being a memory refreshed, a continuing of old experience in new combinations?

But in that case where are the words? In the same place, we would have to say, as the music we heard last night that comes back to us this morning. 'You become the music while the music lasts,' T. S. Eliot said. And perhaps we continue to be so intermittently even afterwards. In that case, the word is always where we last came across it, spread out across our lives. As much 'out there' as anything else we experience.

Let's push this idea a little harder; when we talk about the *mixing* or *superimposition* of experiences in dreams and thought, are we sure that these really constitute special cases? Is experience ever *unmixed*? All recent research on food, for example, tells us that taste changes in relation to sound, to smell, to colour, to texture. The four other senses.

Play crunchy sounds through a loudspeaker while someone is eating cereal and they have a more positive impression. Serve the same product in blue form rather than brown and they have a negative impression. Call it slops and people react one way, call it salmagundi and the reaction changes. At that point mightn't it be logical to say that taste is *inseparable* from the other senses, that it is also made up of seeing and hearing and smelling and touching? And, vice versa, don't things perhaps look different if there is birdsong and romantic music in the background? Isn't the music better if experienced in a certain environment, sung by a beautiful singer perhaps? And so on. It's true one can remove a sense from the mix by disabling the part of the body that allows, say, sight or sound to happen; it's also true that one can choose to focus more on one aspect of an experience than another, the sight rather than the sound, the taste rather than the smell; but, in general, when all the senses are active, how can we ever say that our experience of them, or rather the overall object experienced, is not integrated? Isn't this the logic of calling colours cool, or a girl tasty? Dante speaks of the land where 'the sun is silent'. 'The orchestra was playing yellow cocktail music,' observes Fitzgerald in *The Great Gatsby*. Literature is full of synaesthesia. It seems our bodies allow us to intercept, or carve out, a variety of things from the world and experience them *together*, as a single object of perception. So, when I feel pain in my ankles as I meditate, it may well be fused with the sound of a truck ticking over outside the house, or the field of red light behind closed eyes, due perhaps to the first rays of sunshine penetrating the room. All together these things form a single experience, not unlike the way reflections on a glass display case merge with the products behind to form a single visual perception. At which point it makes sense to say the ache is red, or the pain is ticking over. Or the redness and the ticking over are achy and painful. And this need, or merely desire to resort to synaesthesia, to use apparently nonsensical juxtapositions, perhaps reminds us that it is *only in language* that objects and senses are ever really separate. Only words give us the impression that the world is split up. When we actually experience the world, we get the whole lot together, as when I woke in the morning and saw the silver grey wallpaper, the dark wardrobe, the white sheets, and so on, with no spaces between them and no punctuation.

But back to language and its location. From the earliest age we all had parents and teachers endlessly repeating words to us and soliciting words from us – This is a *tree*, John, say *tree*. That is a *cat*, Sue, say *cat*. We could not produce language if we had not heard it a million times, as we could not dream or hallucinate trees if we had not seen them with our eyes. And perhaps we like to attach words to things because the experience of the thing changes subtly when joined or blended to particular words and their sounds associated with particular past experiences; use a different word for the same thing and you have a different experience, a different synaesthesia. In this scenario, writing and thinking are simply *performances*, like walking or cooking, using items out there in the world, in this case the sounds we have heard and the signs we have seen a thousand times, to cook up something new.

In his book *Supersizing the Mind*, Andy Clark remarks on

the economy of human walking compared with robotic walking. A human being uses far less energy than a robot. Why? Because at each step the robot moves from stillness to motion, activating a power-driven motor, while the human body, leaning forward and with all the subtle jointing of ankle, knee and hip, exploits the combination of firm ground and gravity. Gravity draws us to fall and the leg intervenes to block the fall finding solid ground to push forward into the next potential tumble. We mesh with the environment to walk with little effort, so that if one had to list the causes of our movement, our perambulation, one of them would have to be gravity and another solid ground.

Could we say something similar about the movement of our thinking and speaking in words? Something that helps us understand both the nature of thought and the location of language? We make the initial movement, which is to say we speak a word, at which point the inertia of grammar, syntax and lexical associations draws us forward, because syntax demands that the sentence be completed. Walking, we have a choice of paths and streets ahead; the body sets off, meshing its efforts with topography and gravity. Speaking, there are always a fairly limited selection of places we can go (now that I've finally acquired a smartphone, its dictionary predicts them well enough), while the territory we move on is that of the words we have heard before. It takes a big effort to force the language in some new direction; in general, our perceptual experience moves around, in a *pas de deux* with old words.

No sooner do I write this than I realise that I have simply arrived at something Wittgenstein said decades ago, that there was no reason at all to suppose that language use, even in silent thought, was *internal to the brain*, but rather, like speech, a performance, something you *do*. The curiosity, then, is why so few have taken on board what Wittgenstein so lucidly argued. Perhaps the problem is that we simply do not want to take these ideas on board.

If you ask your vipassana teacher about thinking in words, he will dismiss most of it as 'proliferation'. In general, Buddhist teaching offers a vision of the relation between language and perception which could easily be integrated with either Hume's or Manzotti's positions. Where we in the West have five senses, Buddhism recognises six; 'perceptive gates' they are called. Five of these gates correspond to our five senses. The sixth is the mind. In this view the mind is not an organ that *initiates* thought, but an instrument of sense perception that *intercepts* thoughts, which are assumed to have the status of sense objects, like trees, or apples, or music, or the smell of bananas; and the mind does this – 'thinks' – particularly when the other senses are not active and focused. After which, it tends to 'proliferate', that is, it intercepts thought after thought after thought. Pain and pleasure – which is as much as to say experience – come to us through all six of these perceptive gates, which, in meditation, we seek to close down in order to move beyond pain and pleasure, for a little while. A typical piece of Buddhist teaching reads as follows:

Here let me ask you a question. Which of these six gates makes us suffer most? The mind. How do we know? Through experience. When you begin to meditate your mind is concentrated on your breathing, on the rise and fall of your abdomen. But gradually as your efforts flag, concentration weakens, and then maybe your mind shifts to your son. And if your son is very good you have pleasant feelings about him. But if your son is bad you have unpleasant feelings. The pleasant feelings then cause attachment or desire or love to arise, while unpleasant feelings, or aversion, cause anger or hatred or disgust to arise. Then you suffer. You are sitting here meditating, but you are also suffering. Why? Because you could not close your mind's gate and these thoughts have crept in.

So how am I faring with the mind's gate this early morning in Heidelberg? I have been sitting for quite a while now, though I really have no idea how long. Normally I use a little program on my computer that will give a soft gong sound when an hour is up. The advantage of this is that you can know the time has passed without having to move to turn an alarm off. You can stop or go on as you choose. But here in the hotel I haven't set up the gong for fear of waking Eleonora. And I don't know how much time has passed. Still, I am aware that gradually my perseverance with this long slow body scan has begun to yield fruit. Perception's gates are closing and a new mood is settling in. It is a wordless awareness that seems to fuse physical sensation and attentiveness, an awareness of quiet stillness, you could say, but also of constant flux. It's true that now and then there are, as it were, patternings of possible thought. Whisperings. Murmurings. As if language didn't want to let go even when, on the surface, it already has. It would be interesting to know what neural correlates this strange phenomenon has. But eventually even these whisperings fade, and I reach the point where all the body's different sensations, all the pulsing and dissolving, are gathered up in one steady awareness of stillness and flow.

Have I arrived at this experience because I have finally divested myself of words, or have I shaken off the words because I focused on an experience to which words cannot easily attach themselves? Certainly, in retrospect, I can find no words that describe this meditative state in any precise way. All I can do is throw ideas at it. I could say, for example, that one has an impression of being both present as yourself but also merged into everything that is not yourself, absolutely awake, ready to spring to your feet if necessary, but also pleasantly annulled. Maybe it would be simpler just to fall back on some hackneyed idiom: I am in the zone.

So what is going on here? Is this insight into the true nature of reality? Does it tell me anything about consciousness.

When I initially took up meditation, I was very aware that the pains I had been experiencing, abdominal for the most part, had largely had to do with a denial, or simply negligence, of my body. I was living so much in my thinking, my writing, that I hadn't realised how tense and knotted up my body had become. I hadn't given it attention. In the months that followed, the hours of silent observation of breathing, the body scans and some yoga of the more meditative variety, had a dramatically beneficial effect. As a result, I quickly bought into the idea that mind and body are one. Indeed for some years this became an item of faith. I *was* my body as much as I was my brain as much as I was my mind. Psychosomatic ailments were not exceptions to a normal body/mind split; on the contrary every 'mental state' was somatised in the body and every body state was mental. Crucially, to close oneself in a verbal mind at the expense of the flesh-and-blood body was a major error.

But however intimate mind and body may be, are they really one and the same? Are we our bodies? Enactivists would tend to say yes, and not only. For them mind and body are one and both merge with the world we engage with. Internalists would say no. The mind is one with the brain, the neurons inside the skull, or is produced by it; of course the brain is part of the body and evolved along with it, but other parts of the body can be lost without a sense of self being diminished, while the brain cannot. It may even be, some of the more imaginative of them propose, that one day the mind could be downloaded from the brain onto some other physical platform – a computer for example – thus escaping from the body altogether. For if consciousness 'supervenes' – that word again – on the billions upon billions of interconnections in the brain, then reproducing those connections on some incredibly powerful computer would reproduce your consciousness. In this way they get back to Descartes' separation of soul and body, and his centring existence in thought and consciousness, rather than the world we live in.

Can those of us neither trained in philosophy, nor steeped in neuroscience, use our own experience to push the argument along? It seems obvious that the state of my body influences the state of my mind. And vice versa. If my eyesight fails, perception alters, perhaps I feel depressed. Or again, if I feel depressed perhaps I don't look after my body and it declines. I get fat perhaps. Or being obsessed only with what's on my computer screen I become stooped and constipated. But the same could also be said of the mind's relation with the world. A bright day, or a change in circumstances, puts me in a good mood; falling in love makes everything I experience brighter and better. And vice versa. Thomas Fuchs describes these connections very well in his many essays on mental illness. Of course with mind and body the connection is more intimate and immediate than with mind and world, but it seems axiomatic that the mind is never independent of *either* the world or the body, in both cases it is influenced and influencing at many levels.

The key difference between body and world in their relation to the mind is the fact that the body is both object and *instrument* of perception. I can see my hand, like any other object, I can touch my toes, smell my armpits, taste my blood, hear myself cough, or sing, whatever. But it is also the body that is allowing me to experience the body like this. Perhaps this explains why sensations in parts of the body change when we pay attention to them.

However, none of this makes the body *identical* with the mind, or consciousness. Certainly, if I lose a part of my body, this would alter my experience, but it wouldn't stop me feeling I was fully a person. Already at sixty-one I have lost various capacities I had at twenty. I can't run as fast or think as fast. I don't have the hair I had. And so on. But far from feeling diminished as a mind, I feel enriched by the sheer quantity of past experience which is now the object of my deferred perceptions; I have a past, a karma if you like. Then there are any number of crucial bodily functions that don't seem to have anything to do with me as mind. Montaigne, who was fascinated by the mind/body relationship, is full of such remarks as, 'That sphincter which serves to discharge our stomachs has dilations and contractions proper to itself, independent of our wishes or even opposed to them; so do those members which are destined to discharge the kidneys.' Beckett, who I suspect read Montaigne regularly, offers this gem: 'gas escapes from my fundament on the least pretext . . . One day I counted [my farts]. Three hundred and fifteen farts in nineteen hours, or an average of over sixteen farts an hour. After all it's not excessive. Four farts every fifteen minutes. It's nothing . . . Damn it, I hardly fart at all . . . Extraordinary how mathematics help you to know yourself.'

Extraordinary indeed. Which is to say, not at all. The mind/body identity formula will have to go. I am not my farts.

But then the idea of the mind being only the sum of neural firings also seems suspect, in that those firings are only occurring thanks to the presence of the body and the world. Consciousness is consciousness of the world, not of neural firings, and the world is not to be found in the head. To speak of the mind being identical with the brain, then, seems as hazardous as supposing it identical with the body.

Perhaps the position we could arrive at is that the mind, consciousness, is identical with whatever is the object, or group of objects of perception in the ongoing flow of time. Of course those objects could and almost always would include the body, or rather some part of it, since we are rarely aware of all of the body at once, as we are rarely aware of all of any object, whether a wardrobe or a Beethoven symphony, at once. But it might also include the planet Mars, of which we would now have to say that although it is 50 million miles from the body that allows me to perceive it, is no distance at all from me, since in part it is me – 'me', or my spread mind, being simply everything I experience through immediate or deferred perception at any time. Sights, sounds, textures, tastes, smells, verbal and non-verbal memory, all endlessly unfolding as attention shifts from one element to another. That is mind.

So, sitting on my quilt, eyes closed, still focused on my barely perceptible breathing, what *is* that experience, that object or array of objects that my mind is now identical with? The body is part of it, the room, though unseen, is part of it, with its still air and vague sounds. The rolled towel beneath my backside is part of it. But mostly the object of my focused experience is an infinitely subtle shifting of air back and forth from my lungs across my lips and the now settled wakefulness of the mind contemplating this uninterrupted movement, and indeed its contemplation of that movement. What it *feels* like is a fine, deep stillness and a constant flow. Both together. Perhaps we could say that with the gates of perception, including the mind, mostly closed, all the tension that naturally occurs when the body is focusing on this or that object is attenuating. And perhaps it is precisely that business of shifting between one object and another that gives us a sense of self, the feeling that we are asserting an individuality by 'choosing' what to look at or what to think about. Having largely eliminated any external objects to perceive, and by now steadied – eyes closed, legs crossed – the body that I am simply forced to perceive, so that there are now few changes in my physical state to engage my mind, my sense of self begins to dissolve. At which point you realise the obvious, that *selfhood is tension.* And however wonderful it may be to feel I am Tim Parks, engaged and embattled with the world, it is even more wonderful to let Tim Parks go, and everything else with him.

For a few minutes. I'm never able to sustain this blessed state for more than a few minutes. All at once I feel that it is time to end my meditation. How do I know? I just know. I climb quietly to my feet and slip back into bed. One is always cold after meditation. The body temperature falls. So I cover myself carefully. And the feeling now is one of immense gratitude that I have learned this technique which has made these moments possible. For whereas an hour ago my body was all agitation, now it is all contented calm. Trying not to grow too attached to the sensation, I drift into sleep. My dream diary for that night reads:

> After meditation. Dream I go to an upstairs bathroom where Mother is bent over the sink. I call her and she doesn't reply. I realise she is dead. I go downstairs, shocked. Then decide to go up again and pull her body from the sink, call a doctor and so on. But at the top of the stairs I hear sighing. She must still be alive. As I walk along the landing Mother pops out of a side room,

full of energy but with her face painted white. She is shocked to see me as she didn't know I was in the house.

We are such stuff as dreams are made on, Prospero says. What stuff would that be? Very likely the same stuff as everything else.

Never Personally Killed a Mouse

The hippocampus and entorhinal cortex play a pivotal role in spatial learning and memory. The two forebrain regions are highly interconnected via excitatory pathways. Using optogenetic tools, we identified and characterised long-range g-aminobutyric acid-releasing (GABAergic) neurons that provide a bidirectional hippocampal-entorhinal inhibitory connectivity and preferentially target GABAergic interneurons. Activation of long-range GABAergic axons enhances sub- and suprathreshold rhythmic theta activity of postsynaptic neurons in the target areas.

Part of the logic of my accepting the invitation to talk to scientists and philosophers in Heidelberg was that the promise of these encounters and my determination not to appear ignorant would force me to do the kind of reading I am always putting off, or always beginning and then setting aside, because so arduous. The opening paragraph (above) of the article that Professor Hannah Monyer sent me about her current research is a typical example. What is the entorhinal cortex, what are excitatory pathways? How do you use optogenetic tools and what are they for? What is a long-range g-aminobutyric acid-releasing (GABAergic) neuron when it's at home? And so on. Above all, how much time does a person like myself need to invest to get even a sketchy understanding of what neuroscience is up to and what is at stake anyway? What is being claimed here?

'Study a map of the brain,' Riccardo told me, 'and read any basic textbook on neuroscience.'

'Including *Neuroscience for Dummies*?'

'Why not? It's written by a respectable neuroscientist. It covers the bases.'

So, I studied some brain maps and, after Christof Koch's *Quest for Consciousness*, read *Neuroscience for Dummies* which is written by Franklin Amthor, director of a neuro-science graduate programme in the psychology department of the University of Alabama at Birmingham. The department's website asks the question 'Why major in PSYCHOLOGY?', floating the words over a background picture of the brain, while another picture shows students looking at a screen displaying a diagram of the brain. Psychology *is* the brain. So it seems. Mind is the brain. Consciousness is the brain. Mental illness is the brain. And *Neuroscience for Dummies* might best be described as an extended map of the brain, or account of the topography of the brain where every form of experience and activity is correlated with a specific area of the brain and specific, highly complex behaviour on the part of the neurons and chemicals in these different areas. That is, understanding the brain seems largely to entail figuring out what area does what. Or at least is correlated with what.

Certainly as you read *Neuroscience for Dummies* it's not a bad idea to have a computer handy with the famous Allen Institute's remarkable Brainspan website open for consultation. This wonderful facility will offer you simply endless diagrams, scans and images of every nook and cranny, every hole and corner, of this bizarre organ. Each time some new area with some different function is mentioned, you can go and see where it is.

But don't imagine you are soon going to find your way around. Exploring the brain is not like examining the map of a new town where you can check street names on paper with signs at a corner, or even an anatomical diagram of the human body, where the various parts can easily be related to our own hands and legs, torso and organs, etc. Of course, we don't see our own organs, except perhaps during an ultrasound scan, but they have all made themselves present from time to time. We are aware of an overloaded stomach, aware of wheezy lungs, a thumping heart, kidney pains, bladder aches, and so on. So we have a fair idea of where these organs must be. But we have no experience of separate parts of the brain. We have been told that the right hemisphere controls the left half of the body and vice versa, but when we take a screwdriver in our right hand we do not feel anything in the left side of our heads. Nor do we feel the left side of the brain when we talk or read, despite the fact that it is apparently this hemisphere that is responsible for language. Of course we say hemisphere knowing full well that the brain is not spherical. And when we look at images of the brain, they are rarely real photographs, but neat and idealised diagrams. Because a photograph, when you find one, just shows a gruesome pinkish grey, vaguely intestinal lump.

All this simply to say that nothing prepares us for the complexity and frankly, oddity, of the brain's anatomy. It may be only about a kilo and a half in weight, it may have a modest volume (in the range of 1,100 to 1,300 cc), but it is a tantalisingly amorphous, densely packed conglomeration of maddeningly disparate and quirkily shaped parts. If you can soon enough learn to recognise up and down, outer cortex and inner cerebrum, the frontal lobe high up at the front and the cerebellum low down at the back with the occipital and parietal lobes under the scalp in between and the temporal lobe squeezed in the middle (always in neat textbook illustrations of course), you are going to be hard-pressed indeed to identify the hundred and more smaller regions and subregions and glands, all of quite different sizes and shapes, squeezing puzzle-like into the separate hemispheres, to the point that you have to wonder about the decades if not centuries of effort that have gone into identifying them, not to mention the vexed question of how exactly they are at once separated from each other and connected to each other in such a compact muddy mass.

Even should you memorise these various locations, nothing will help you correlate the exotic names of the various parts – amygdala, lamina quadrigemina, cingulate gyri – with their sizes, locations and functions. It seems particularly perverse, for example, that it should be the occipital lobe at the bottom rear of the brain, about as far away as possible from the eyes, that is chiefly responsible for vision, while the orbitofrontal, piriform and insula areas responsible for smell are right behind the eyes; one says 'responsible', without really knowing what the word means in this context, indeed without any real sense of what exactly is going on. For of course a whole host of other elements are involved in vision, not least the object seen, the light reflected on it, and the fantastically complex anatomy and chemistry of the eye.

Some very small brain areas seem to have an extraordinary array of tasks. 'The hypothalamus', we are told on one medical website, 'is located in the floor of the third ventricle and is the master control of the autonomic system. It plays a role in controlling behaviours such as hunger, thirst, sleep, and sexual response. It also regulates body temperature, blood pressure, emotions, and secretion of hormones.' These are big jobs. 'In humans, the hypothalamus is the size of an almond,' we then discover.

But not all of the brain is working so diligently. 'Neuroscientists at the University of Ingberg', another site tells us, 'have found a brain region that does absolutely nothing. Presented at the annual meeting of the Society for Neuroscience, their research showed that a small region of the cortex located near the posterior section of the cingulate gyrus responded to "not one of our 46 experimental manipulations".'

Something you can easily hold on to when nothing seems to make sense is the idea that there are older and newer areas of the brain. In evolutionary terms of course. So the most basic and involuntary activities – breathing, swallowing and digesting, defecating, vomiting and sneezing – would appear to be 'governed' by various neuronal clusters – the pons and medulla oblongata – at the very core of the brain, around the top of the spinal chord, structures shared by even the most 'primitive' of animals, while other more sophisticated activities are associated with areas that were, so to speak, added on around these. Still close to the core is the so-called limbic system – the hippocampus, the amygdalae, the septal nuclei and others – a tightly packed assortment of parts whose activities can be correlated with memory, emotions, the sense of space, pleasure, and many other heterogeneous activities.

Around and on top of the limbic system is the catch-all cerebrum which gathers together the various 'lobes'. While the deeper parts of the brain appear as dense clusters, the lobes take the form of layers, folding over on themselves to give that familiar wrinkled surface we see in iconic images of the brain. These 'newer' areas are apparently responsible for the higher cognitive functions, language, thought and 'related forms of information processing', though intriguingly, as the Mayfield Neurological Surgery clinic website laconically observes, 'most of the cerebrum's activity is subconscious'.

What makes this old–new idea easy to grasp of course is that it proposes a narrative – we love narratives – and within the narrative a flattering hierarchy. In the beginning there were only the basic activities that any amoeba can manage, then bit by bit, climbing up the evolutionary tree, we added on all the things that, at the last, make us humans unique and special. So the brain becomes an objective correlative of progress and human superiority. This is rather comforting, even if, in more serious accounts, the formulae 'may have', 'might have', 'could have' and 'would seem to' abound.

Never mind. Introducing *Neuroscience for Dummies*, Franklin Amthor is bullish. 'I believe we can understand how the brain makes us what we are,' he tells us. Essentially, it is composed 'of neurons, each of which is a complex little computer. Parts of the nervous system make suggestions to the rest of it about what you should do next. Other parts process the sensory inputs you receive and tell the system how things are going so far. Still other parts, particularly those associated with language, make up a running dialog about all of this as it is going on; this is your consciousness.'

Of course, in the normal way one tends to zip through opening paragraphs like this and hurry on. But if one stops and tries to get a sense of the overall picture that Amthor takes as his departure point, it does seem puzzling. Puzzling that our brains are made up of things – computers – that we ourselves only recently invented. Puzzling that the nervous system can be divided into parts that, like people, appear to possess intention and volition, suggesting and telling each other things (are we back to the little fellows in Pixar's *Inside Out?*). Puzzling also that there are parts 'associated with language' without our knowing in what way associated, or to what end, and that these parts then make up a dialogue (between each other?) about all this. Puzzling too that he takes for granted that it is the brain that 'makes us what we are'. Is there nothing else to being human?

Amthor remains confident. 'Those concepts aren't too difficult to grasp,' he insists, 'but people think of neuroscience as hard. And why? Because in order for your nervous system to perform these functions, it takes 100 billion neurons and a quadrillion connections structured over billions of years of evolution and all the human years of development and learning that resulted in who you are and where you are now.'

Paradoxically, this seems to me altogether easier than the previous paragraph. It's easier, that is, to grasp the formidable nature of the neuroscientist's task. Figures like 100 billion and then a quadrillion can only be grand approximations (more recent counts suggest around 85 billion neurons) but they do bring home the single most extraordinary fact about the brain: it is a monster of connectivity. Varied as they are, each of those endless neurons has hundreds or even thousands of dendrites, tiny filaments that bush out branch- and twig-like from the cell body to meet the dendrites of other cells, and each has an axon, a single, more robust filament which may be as much as a metre long, stretching out into the tangled meanders of the brain to connect with distant dendrites or other axons. All these dendrites and axons can carry and receive electrical charges, which may fire off at different frequencies, in unison or in opposition. Between these connecting filaments are the so-called synapses, sealed junctions where, in response to an arriving electrical charge from one side, chemical substances of various kinds are released and thus alter the state of the dendrite or axon on the other side, causing a new electrical charge to travel to the receiving (postsynaptic) neuron which can then react to it by activating, or not, other charges along its axon or dendrites. And so on. Some of these combinations of charge and chemical reaction can be excitatory, that is, prompting more response from the next neuron in the line, some can be inhibitory, blocking activity already present. For every neuron there are about 10,000 synapses. A lot of impulses are going in and out of the same tiny cell.

Since it is received wisdom that the brain functions 'as a computer', all this electrical and chemical activity tends to be referred to as 'information'. 'As information processors,' Amthor tells us, 'neurons receive information from other neurons, perform computations on that information, and send the output of those computations to other neurons.' However, given that no semantic content is observable in any of these tiny charges – but how could it be? – and given again that the charges activated in, for example, an area that correlates with vision are in no way distinguishable from those correlated with touch, or indeed with running a marathon ('Each action potential pulse [the electrical charges running through axons] is essentially identical,' Amthor tells us), it is generally assumed that the 'information' is 'encoded'.

Here we run up against a considerable conceptual problem. A code, the Merriam-Webster dictionary tells us, is 'a system of signals or symbols for communication'. It is made up for a purpose. It has rules. Users on both sides of any message know these rules and know what it means both to encode and decode. Hence, regardless of who sends or receives a given message, its content remains the same.

It is honestly hard to see how such an idea can be applied to our billions of neurons. If, as we said, the charges whizzing around in an area of the brain 'associated with language' are 'essentially identical' to those zinging about in an area apparently responsible for smell, do we have to suppose that the meaning of messages changes according to who is sending and receiving them, that there is a fresh code for any possible pair of neurons, or groups of similarly dedicated neurons 'communicating'? This is the stance Amthor appears to take:

> The meaning of the activity (or excitement level) of a neuron is a function of who is talking to it and who it is talking to. If our particular stockbroker neuron, for example, were listening to the output of other neurons talking about things like electrical wiring, certain mines in Chile, and production of pennies, our stockbroker neuron might embody information about the price of copper.

Charming as this is, you have to wonder how the neuron knows it is a stockbroker neuron, knows it is playing the market, knows its companions are talking about mines in Chile, when the market is in no way represented in it or in the electrical impulses going to and fro, but only exists *outside the head.* Like everything else we experience. The microchips in a computer are constantly performing computations on data without themselves knowing the significance of what they are doing and without there being any difference in the kind of electrical activity going on, whatever the content; but this is hardly a problem since the human beings who designed the computer, and who wrote the software/ code and fed in the data, do know. They can interpret the readout. The same intelligence is there at the beginning and at the end of what is a planned process. This is just not the case with the neurons in the brain.

Back in 1948, Claude Shannon, who formulated the mathematical theory of information that would be so important for the development of computing, remarked that 'The fundamental problem of communication is that of reproducing at one point either exactly or approximately a message selected at another point.' Information could thus be expressed as a probability that the message sent and the message received were identical. For this to occur you needed a transmitter, a receiver and an agreed code. It makes sense. In later publications, Shannon observed that since none of these elements were present in the brain, it really couldn't be described as an information processor. Hence the analogy that Amthor, and with him almost all neuroscientists, use to describe what is going on in the brain, is shaky.

What to do?

Perhaps, to purge our description of potentially misleading narratives, it would be better simply to say that a truly astonishing number of interrelated *changes* are constantly and simultaneously occurring throughout the brain in apparent relation to things happening outside the body or in conjunction with what the body itself is doing, or again, in relation to no experience or activity that we can identify, either within the body or without, since all recent experiments suggest that our busy neurons continue to exchange their charges and chemicals even when we are doing nothing and thinking of nothing, eyes closed and lying down.

In any event, it is this extraordinary intensity of cerebral connectivity, however configured or interpreted, that has led some scientists and philosophers, in particular Giulio Tononi and more recently Christof Koch, to suppose that if you built a computer with the same number of connections and the same intensity of internal 'communication', it would somehow become conscious. Consciousness, that is, would naturally, or magically, 'supervene' on it. In short, you take the most extraordinary fact about the brain – connectivity – and by reproducing that fact hope you will arrive at the wonderful phenomenon you have assumed, but in no way proved, that the brain is responsible for: consciousness.

Let us now add to all our mind-boggling numbers the fact that beside the 85 billion neurons in the brain there are also something between 1,000 and 5,000 billion glia cells. Yes, the numbers are that vague. 'About 10–50 times as many glia cells as neurons.' Again, the literature here suggests a hierarchy, with the glia cells – the word meaning 'glue' in Greek – holding together the neurons and coating their axons with myelin, a fatty substance that insulates them against the dispersal of their electrical charges, thus allowing for faster and more powerful electrical charges. They are helpers, servants, not the real thing. What matters is the computer-like electrical impulses, not the surrounding insulation. However, despite this apparently humble role it turns out that there are also all kinds of chemical reactions going on among these glia cells which can change the shape or structure of the neurons. So in the end it seems unclear that they are merely facilitators.

Frank Amthor is imperturbable. 'You need to know three things to understand how the nervous system works,' he winds up his preamble. 'The first is how the neurons themselves work. The second is how neurons talk to each other in neural circuits. The third is how neural circuits form a particular set of functional modules in the brain. The particular set of modules that you have make you human. The content of your specific modules make you, you.'

Functional modules? Content thereof? Isn't the whole problem of brain science the fact that it is impossible to identify *content* in the brain?

'Perhaps,' Eleonora suggests, when I read this out to her, 'neuroscience really is for dummies.'

Is she teasing me, I wonder, or Amthor? Or perhaps anyone mad enough to take on quadrillions of synapses?

Over our second breakfast in our Heidelberg hotel, where today our lady of the yellow hat is dressed all in sky blue, I feel like a student heading for an exam he is bound to fail.

'You've read a stack of books,' Eleonora says sympathetically. And this is true. 'And endless academic papers. Including Monyer's.'

When I sigh, she says, 'At least you have hot tea this morning.'

'Learning occurs,' Frank Amthor says, 'when experiences modify the strength and identity of the interconnections between neurons and thus create memory.'

How this modification takes place, whether it can be observed and measured, what is meant by the 'identity of the interconnections between neurons' and how they might constitute memory, he does not say.

But I had no trouble at all with the tea urn.

*

'It's all much easier than you think,' Hannah Monyer tells me and my first thought is that she is used to reassuring people who are intimidated by what she does. She breezes into the waiting room where I have been sitting for some minutes, apologises for being late, mentions a family problem, smiles, asks if I read her article, tells me not to worry about all the awful jargon and assures me that in a very short time I will understand. She's handsome, very feminine despite a white lab coat and very wired up in a way that I immediately appreciate is not German. In fact, she's Romanian. Her parents brought her to Germany as a child and she's been extremely successful. She studied music and medicine, did her doctoral thesis on the phenomenology of jealousy in the work of Marcel Proust and the psychiatric literature of his time. She laughs, glad to assure me that she is no alien to humanistic studies. On the contrary. She loves novels. Only later did she get into neuroscience. Studied at Stanford, and so on.

'And, in 2004, you won a huge and prestigious prize,' I remind her. The Gottfried Wilhelm Leibniz Prize; 1.5 million euros. 'I watched the YouTube video of you receiving it, though I didn't understand very much, and maybe wouldn't have, even if it had been in English.'

Again she laughs. 'The money was to pay for research of course. It enabled me to take on some excellent young researchers. You can get more done. Though then you have the responsibility of constantly finding more funding for them. A lot of time spent filling in applications, explaining what you hope to achieve. This kind of work is expensive.'

Certainly, the facility I am in is altogether more moneyed

and purposeful than the genteel and creaky rooms where Sabina Pauen and Thomas Fuchs do their thinking. We're in the Deutsches Krebsforschungszentrum, a glistening, aseptic, white and grey structure with lots of glass and right angles, dedicated to cancer research. Monyer's work has also won her the smaller Philip Morris Prize for outstanding research – I have decided not to quiz her about the paradox of someone under the cancer-research umbrella receiving prizes from tobacco companies – and in 2010 was awarded 1.87 million euros by the European Research Council. Clearly society sets a far higher value on the work she is doing than on research in other fields. So what is it exactly that she is discovering?

'What did you understand from the article?' she asks me, brushing back her long hair. 'So I know where to start.'

'Well,' I take a deep breath. 'You are looking at exactly what goes on, chemically, at the synapses between the neurons, particularly in the hippocampus, which is an area in the evolutionarily older part of the brain that is active when we deal with issues of space and memory.'

'Perfect,' she cries. As if these remarks themselves were worthy of a prize.

'Really, I'm just parroting,' I tell her. 'I haven't actually *understood* anything.'

'But you have! Go on.'

'In particular you're studying interneurons, a special kind of neuron, which helps link sensory neurons, presumably busy with perception, and motor neurons, presumably busy with the body's actions, into related circuits, sometimes calling together quite distant parts of the brain, again presumably so that what is perceived on one side can then elicit an appropriate response on the other. *GABAergic interneurons* to be specific, though I have no idea what they are.'

'Horrible jargon!' she agrees.

'And glutamate receptors.'

'Right.'

'Which again, I don't understand, though I know they have to do with these extremely delicate exchanges that take place where the dendrites and axons of these neurons meet.'

'It's part of the chemistry of the receptors at the synapses.'

'What else? You work mainly with mice and have built up a complicated and, Wikipedia tells me, innovative bag of tricks for teasing info out of them.'

'Excellent.' She even claps her hands. 'Essentially, we're just trying to understand how the neurons communicate and how information gets passed around and stored in the brain, mapping out where these interneurons connect, often much further away than was previously imagined. The brain in general seems more linked up than people originally thought.'

Information. Storage. I decide to wait before we talk about this.

She frowns. 'But I was told that you were interested in consciousness, whereas to be honest my work is mainly involved with memory.'

'A lot of our consciousness seems to be made up of memory.'

'Of course.' She jumps to her feet. There is a charming *Sound of Music* enthusiasm about the woman, as if she were about to burst into song. 'Let me show you an experiment, that's the easiest thing.'

It's interesting that with all three of the people I've met to date, you can sense at once that they enjoy an enviable purposefulness. They are focused, engaged, active, and they live inside worlds where everything *means* intensely. At least to them. In Thomas Fuchs's enactivist vision of consciousness, they are all healthy people.

What feels much less healthy is the smell in the room I am now shown into. It is a small space that seems part office, part lab and is characterised above all by a very strong smell of mice. The sense of smell, we remember, from Galileo on, is reputedly generated entirely in the head, it doesn't actually exist in the world, yet my head certainly hadn't been producing this phenomenon just a few moments ago. Nor is it enjoying what it is producing now. The mouse odour is not quite a stench, but getting there, and along with it, instant and unbidden, comes a memory from perhaps fifty years ago of the time when I used to keep mice myself. A pet shop in North Finchley had sold me two innocent white mice in a small wooden cage. Originally, I had meant to keep them in my room, but very soon the smell began to bother me, so that after some negotiation with my parents I moved them into the outside boiler room of our sprawling Victorian vicarage. Here, in no more than a few months, the original pair had multiplied dramatically, gnawed their way out of the cage and were scampering in their dozens in and under the synthetic lagging around the house's ancient oil-fired boiler.

Hannah Monyer points to a single white mouse in a deep metal box on a low table and begins to describe what they are doing with it. But even as she does so I am aware that if I wanted to I could explore and expand this old memory of mine at will, building it up from the trigger of the mousey smell to include the wooden workbench to the left of the boiler-room door with its scatter of rusty tools, the roar when the boiler fired, the mouse droppings on every exposed surface, and so on. Memory, perhaps, it occurs to me in this lab where memory is being studied, is a routine you find yourself *performing* in response to a prompt, the way some music has you hazarding a dance step or two almost against your will. It's something that both binds and enriches. But I must listen to what Professor Monyer is telling me. Because it's complicated.

Unlike the mice I kept, this creature we're looking at in the Deutsches Krebsforschungszentrum, and indeed the four or five other similar animals in cages stacked against the wall, has a tiny black plastic box attached to the top of its head. A little less than a centimetre across and sitting slightly to the side of the head, this unexpected contrivance has the air of some incongruous military hat and bestows on the mouse an even more comic and pathetic look than little white mice ordinarily have. The creature looks vulnerable and ill at ease, crouched in his cage with his black box on his head. Or maybe her head.

'Protruding from under the box are four electrodes,' Monyer tells me, 'much finer than needles, of course, that pass through the skull and the upper brain into the hippocampus.'

For a moment, staring at the mouse, I wonder at the practicalities of this. The animal's head is so small. The hippocampus is not a large part of the brain, nor is it near the surface.

'Can you really be sure that the electrodes have gone where they're supposed to?'

Monyer is breezy. Her lab technician is brilliant, she assures me. He has long experience. Above all, the hippocampus has a characteristic firing pattern, that is to say, its neurons send their electrical impulses at recognisable intervals. 'So when the electrodes register that pattern, we know they're in the right place.'

'But aren't they painful? For the mouse?'

'Not at all.'

I'm reminded that while the brain may be 'responsible' for the pain we feel in other parts of the body, it is apparently immune to pain itself. You don't feel a scalpel cutting into it. So they say.

'And?'

'As the mouse moves around the cage, the electrodes record the activity of the nearby neurons, or hopefully, if we've set things up correctly and we're lucky, of a single nearby neuron.'

The cage is about 40 by 40 cm and now I notice that there is a video camera placed directly above, presumably tracking the movements of the mouse.

'With the electrodes and the camera synchronised, we can associate the movement of the mouse around the cage with the firing of these neurons. For example, we've noticed that when the mouse has its left side against the near wall, leaving the rest of the cage empty to its right, then a particular neuron close by one of the electrodes always fires off, sends out an electrical signal.'

'Can you really be sure it's the same neuron?'

One book I have read claims that about a million neurons could fit into a grain of rice.

'The signal is always received by the same electrode and is always the same strength. We're confident it's the same.'

'And this means?'

'The neuron is telling the mouse where it is in relation to the walls of the cage,' Monyer says. There are also neurons that fire when the mouse is at any part of the border of the cage, others again when it is at any one of four grid-related positions that form a rhomboid in relation to the square of the cage. So the creature would seem to have a kind of map in its head.

'Surely,' I object, 'it is not *telling* the mouse anything. The mouse isn't separate from its neurons, is it?'

'Of course it's a metaphorical way of speaking,' Monyer acknowledges. 'Let's say the neuron is receiving input, processing it and passing it on.'

But where to? And why?

For a few days the research team tracks the mouse's movements and neural activity in the cage. And here's a curious thing: when the mouse rests or sleeps, they notice the same neurons fire in the same sequences, or in curious reversed sequences, as they did when the mouse was moving around – but faster. Ten to twenty times faster. Monyer shows me printouts where these firing sequences have been expressed as graphs. You can compare the wakeful versions with the sleep versions. The patterns are not quite the same, but clearly similar, or symmetrically inverted. It's as if, in sleep, the mouse were reliving its experience, in fast forward and fast rewind, and committing it to memory in some way. So that then when the experience is repeated – when, that is, the mouse finds itself in the same position in the cage (which must be rather often) – perception and memory are simultaneous as the familiar pattern fires off. At which point memory could be understood not exactly as something stored, but as a tendency for neural patterns to repeat. A kind of habit. Perhaps. Though why the repetition would come with a conscious recollection that it is a repetition is not clear. In any event, these are the stories one tells oneself in response to the read–outs produced after inserting electrodes into a mouse's hippocampus.

After a few days, when enough data has been collected – that is, when nothing new is being observed – the mouse is decapitated and its brain instantly sliced into super-thin layers whose cells will live on for a few hours despite the animal's death.

I suppose it's pointless asking at this point whether the mouse suffers or whether it matters in any way at whatever level that a creature was alive one moment and dead the next. This is serious science of the kind that could one day bring benefits to human beings with distressing brain conditions. All the same, and ridiculous as it may seem, there's a part of me that is not altogether on board with these decapitations, even if, as it turns out, the mouse is anaesthetised first. Nor would I like this uneasiness to go away.

With the skull peeled off then and the brain sliced as thinly as anything can be sliced, and while the cells are *still living*, a powerful, interactive microscope is used to focus on single neurons and even open them up. By introducing an electrical stimulation, it is possible to trace their axons (long hair-like tendrils) and see where each cell connects, though only of course on the horizontal plane of the wafer of brain that has been cut. Certain neurons in the hippocampus have axons that reach far out into the cortex, where information is 'processed and stored as long-term memory'.

'Stored? The information is actually in there?'

'Let's just say neurons fire in that area when we are remembering things. What's important is that we have established some neural correlates of consciousness. This neural activity takes place while this experience occurs. These are empirical facts.'

The hallmark, I suppose, of an empty notion is that you can set it aside without its making any difference. 'Stored' and 'information' are simply ditched.

Having prepared me for what I am about to see, Monyer now leads me into the lab proper, which is just the other side of the corridor from her office. Again I should say that what I remember of this interview and visit, writing almost a year after it took place, and with only scanty notes about the salient moments of our conversation, is the arrangement of the space and my position in it. Which means, above all, my position in relation to Monyer and the three young women working in there. So I recall that I was taken through a nar-row entrance way with a screen to my left and a wall to my right, after which we turned left and walked between machines and workstations on every side to the far end of the lab where the mice brains were extracted from their skulls and prepared for slicing. No doubt it would have been an interesting exercise to record the neural firings in my hippocampus as I made that walk across the lab and again as I write about it now and then perhaps a third time when I reread what I have written, snipping a word here and adding another there, recalling that there were two young women seated at workstations to our left and one to our right as we crossed the lab. But even assuming this were done and the results showed satisfyingly similar patterns, I'm not sure this would mean that the memory was *stored* in the firing pattern. Might it not equally suggest that my finding my way back to the fact, leaning on one fact after another, as I have been doing throughout this book, in much the same way as one seeks out an old path in the fog, has caused those neurons to fire? They are one with the memory, perhaps I can't have the memory without their firing, but they are not identical to the memory itself, which is more as though, from a great distance, uncertainly glimpsed and much dis-torted by the rarefied air between, the lab were still having an effect on me, still present to me.

But back to decapitations. First the mouse brain is dis-sected using a moderately powerful stereo microscope to isolate exactly the area we want to slice. Then the desired chunk of fresh tissue is set in a small block of transparent gel which protects the cells and makes it easier to slice them.

After being cooled to harden a little, in the fridge, the block of gel is then glued to a specimen holder which, with ice packed all around to keep the tissue alive, is inserted in a charming shiny white machine called a vibrating microtome that looks like an oversized kitchen appliance. Armed with a traditional razorblade this hi-tech machine will cut the tissue into slices as thin as 0.1 mm.

You might wonder why I am bothering you with these technical details – though to be honest I could have added many more – rather than going straight to the results and the reflections they give rise to. The fact is that this meeting of organic tissue and hi-tech machine has its special fascination. The moment you step into the lab, and see the researchers, all women in this case, ministering patiently and silently to their expensive equipment, among batteries of switches, computer screens, keypad control consoles, binocular-like viewing apparatus, etc., the moment you smell that unforgettable blend of mouse and polymers, hear the low, hi-tech hum, you are aware you have stepped into a place of enchantment, a place for acolytes and initiates. There is an atmosphere of ritual and profound mutual respect among those who know, those who are penetrating the heart of life, delving into the very neurons. It's true the smoky candlelight of olden times has been replaced by an intense, dead-white fluorescence (there are no windows in the lab) but the sense of being removed from the ordinary world in order to serve some higher purpose is very much the same.

Later, seeking to remember more clearly what I had seen in Monyer's lab, or to make better sense of what I remembered, I found any number of YouTube videos, whether promotional films or teaching material, in which these techniques for examining live brain cells – almost always mouse cells – were diligently illustrated. And despite the amateurishness and the YouTubeness, the same atmosphere of devotion was unmistakeably there. These people are absolutely and understandably fascinated and buoyed up by the extraordinary complexity of what they are doing and the level of know-how, practical and intellectual, necessary for making their experiments work. And this I believe has important consequences for what gets communicated at the end of the day to people like ourselves who will never spend months and years of our lives obsessed with the logistics of recording a GABAergic neuron in action.

Examining the microtome and briefly contemplating a mouse brain in gel, which looked uncomfortably like a scrap of roadkill, I asked Monyer whether she herself ever got involved in decapitating and brain-slicing and she frowned and said no she didn't. It required, she said, highly skilled personnel. But it seemed to me as she spoke, that there was also a faint intimation of resistance on her part, of not wishing to contemplate that side of the business, which immediately made her more interesting to me. We then turned back to a workspace more or less in the middle of the lab where a young woman was in the process of examining the brain slices.

Monyer introduced me to this assistant in generous terms as a literature professor and novelist. Someone who deserved respect. But also someone who was not one of us, and hence at best could only take away the most superficial notions of what was going on. The woman, in her mid-twenties, a PhD student, was bright, engaging and attractive, like so many of my own students in Milan. And as she worked – and she clearly didn't want to stop working, since the cells in the slices will only be alive for a certain number of hours – she set about describing her equipment. At the centre of it all is a magnificent microscope for electrophysiology, about a metre high with a binocular viewer and all kinds of accessories, workspaces, control pads, drip feeds, keyboards, computer screens and the like. The brain slice, which now has the aspect of faintly mottled tissue paper, is placed in a tiny bath in a specimen dish where it can be constantly irrigated throughout the whole process. Since any vibration at all will disturb the experiment, once the specimen is mounted, the researcher will never touch the microscope, but operate it remotely, viewing the procedure on a computer screen. Meantime the microscope is capable of substituting its objectives – the pieces holding the lenses – for greater and greater magnification, up to 1,000x, as the researcher homes in on the cell they want to investigate, without the slightest vibration. When a suitable cell has been identified, a remotely controlled mechanical arm lowers a pipette so that it just touches the outer membrane.

This is not the kind of pipette we used to use in chemistry lessons. Its point has been narrowed to a diameter of a single micron, or a millionth of a metre. This in itself seems an extraordinary achievement. Once this tiny aperture is in contact with a patch of cell membrane, the researcher can apply a little suction, thus clamping the patch to the pipette, and if required, with a little more suction, break and open the membrane, but only inside the patch, so that any exchange between neuron and outside world is contained in the pipette. At this point all kinds of other operations can be carried out, depending on the equipment in the pipette being used. The cell's electrical activity can be recorded, or alternatively it can be electrically stimulated, or the molecules in it can be sucked out and examined, or fluorescent dyes can be injected into it which will then show up in the cell's axon and dendrites. And so on.

Standing beside this young woman while Monyer went off to make a couple of phone calls, I watched while she manoeuvred her pipette to contact a cell and engage with it. This didn't always work smoothly, she explained. You could easily damage the cell, whereas they wanted to record how it behaved when it was healthy. Moving the controls with practised hands, she was concentrated but relaxed. On the screen, the vaguely aquatic images swimming around meant very little to me. 'Because we are so close up,' she explained, 'you can't get the familiar shape of the whole thing.' It was all very delicate, she added, and if you made a mistake you could lose an awful lot of time. When I asked her what the activity she would be observing actually meant, what was at stake, she smiled and said that what they were after was the interplay of chemical and electrical exchange in this part of the mouse's brain, and particularly the length of the neural axons and where they connected.

'The microscope allows you to observe single ion channels of electrical exchange across the membrane surface. And to map reactions to an electrical stimulus applied in the neuron.'

Like a thousand visitors before me no doubt, I had an acute feeling of being both admitted to and excluded from a world that requires the dedication of a lifetime.

At another workstation a second young woman was studying the brains of mice that had been bred on purpose to suffer from early-onset Alzheimer's. Readers who are curious can search 'knockout mice' on Google, where they will find the websites of companies who breed, or 'knock out', mice with particular genetic disorders precisely for this kind of study. 'We provide off-the-shelf constitutive and conditional Knockout mice for over 2,000 genes,' advertises one site. 'Among these, more than 40 possess an Alzheimer's-related phenotype.'

'What we're trying to do,' the young woman said, 'is to learn more about the progress of Alzheimer's.' This was easier, she explained, with the early-onset variety since it kicked in more abruptly and reliably. I asked her if there were proven connections between the early-onset version of the disease and its more normal occurrence, or indeed between mouse Alzheimer's and human Alzheimer's and she gave a wry smile and told me there was still a lot of work to be done on that.

'I'd like you to tell me,' I asked, when Monyer and I were back in our original waiting room for a chat about what we'd seen, 'exactly what you understand by the word information.'

She was surprised. In what sense?

'Well, I've got some idea, I suppose, of the kind of work you're doing in the lab, the electrical impulses, the incredibly complicated architecture of each neuron and then the brain as a whole, but I don't see how this connects very clearly with memory or consciousness. And it seems to me that the word that is trying to connect the two, or that assumes the two are indeed connected, is "information". Information as electrical impulse, information as conscious knowledge, where the mouse is in the cage. The word is bridging a gap. But what exactly do we mean by it?'

She hesitated. 'In this particular context we're talking about the exchanges between neurons.'

'Earlier you suggested that in response to a situation outside there were neural firings inside which in some way communicated, or represented, the situation outside, storing it in some way.'

She nodded.

'But is a response to a situation outside necessarily communication, representation, storage?'

Then I put the cards on the table.

'You see I've been reading people like Christof Koch who claim that all experience occurs inside the brain, that consciousness arises from this intense connectivity of our neurons, and I'm just not convinced that that's the case. It seems such a leap, from the neurons to the experiences we have.'

She seemed at once pleased that this was out in the open, and impatient to have the issue behind us. 'People like Christof Koch,' she said firmly, 'who try to sell these big ideas to the public without themselves having any proper training in philosophy or any real grasp of the concepts involved, just give neuroscience a bad name among serious intellectuals.' She herself, she went on, and it was clear she was rehearsing things she had said many times before, was just not concerned with the larger questions. She was interested in them of course, as a human being, but her research was entirely and unashamedly focused on the chemical realities in the brain. People like Koch, she said, rarely discussed the fact that the brain was not predominantly made up of neurons, but of chemical substances, and that was because we were still largely ignorant of how they worked and interacted and because they didn't necessarily fit in with the electrical signals/computer circuit analogy so favoured by people like Koch. Her supreme and only ambition, she insisted, was to track down to her satisfaction how a single percept – a sense perception – was relayed through the brain and laid down as a memory. The larger philosophical questions simply did not interest her professionally.

As which point I'm afraid I bottled it. In part because I sensed that she did not want to go any further with this discussion – she wanted to talk to me about synapses and synchronisation between behaviour and neurons – and in part because it was anyway very close to lunchtime when Jakob Köllhofer, head of the Deutsch-Amerikanisches Institut, the man who had offered me a considerable honorarium to reflect on the ways science might be replacing religion in people's minds, was coming to pick us up and drive us to a restaurant. What I should have pointed out of course was that to talk about percepts being relayed through the brain, and of memories being 'laid down', already suggested a conceptual model which was evidently internalist and really hardly distinguishable from Koch's, albeit without the dogmatism. I should have pointed out that going back to the seventeenth century there has always been intense debate as to whether empiricism can ever carry out its research without being 'contaminated' by some preconceived idea of what it is looking for. Alva Noë offers the example of David Hubel and Torsten Wiesel, who were awarded the Nobel Prize in 1981 for their work on the neurophysiology of vision, work, Hubel claimed, that was absolutely independent of any guiding hypotheses. But as Noë points out, elsewhere Hubel spoke of following an 'obvious' strategy. That is, of extending previous work on the brain to 'record geniculate cells and cortical cells, map receptive fields with small spots, and look for any further processing of visual information'. In short, Hubel already believed that vision was a matter of the internal brain's processing of information from outside. It took place in the head. To get their results Hubel and Wiesel worked on animals that were 'anesthetized, paralyzed, on artificial respiration; stimuli were presented to eyes whose lids were peeled back and held open with clips; eyes were kept moist and clear by means of contact lenses. It is only', Noë concludes, 'the assumption that vision is something that happens passively inside the brain that could justify conducting research of this sort.'

So I should have questioned Monyer's claim that there was no model of memory and consciousness guiding her work, no philosophical dimension. Not because I want to invalidate what she is doing – far from it, it's absolutely fascinating – but to wonder if there mightn't be other ways the results could be considered. And thinking back now on my meeting with her, and in particular the atmosphere that reigned in the laboratory itself, the understandable dominance of *techne* over significance, I can't help feeling that there is something going on with neuroscience that almost guarantees this drastic gap between the scientific work itself – the meticulous recording, measuring and analysing of chemical processes and electrical activities – and the crude larger narrative in which these findings are presented. The sheer complexity of the brain demands that those studying it be utterly immersed in its labyrinthine meanderings and guarantees that it will remain an object of study and very likely mystery for centuries to come. It is a place where a researcher can set up home, as it were, for the duration. Nobody is going to figure it out any time soon. Nobody is going to be asked to move on. Nobody is going to remove the funding. All the same, some report needs to be given to the world of what is being achieved or where the research is going. At the medical level, real progress can occasionally be made, simply by establishing the organic difference between the normal brain and the diseased brain in order to get a sense of how the latter might be returned to the former. And if we were talking about research into any other part of the body, this might be enough. But when it comes to the brain, people expect more, and neuroscientists are tempted to give us more. And what they give us, if only implicitly, is the notion that the brain is the be all and end all, that all our experience is internal to the brain and everything that we are is essentially a matter of what goes on in those three pounds of grey jelly. 'All of your sensory experiences', neuroscientist David Eagleman tells us, 'are taking place in storms of activity within the computational material of your brain.' It's hard not to note here, in passing, the desire to dramatise in 'storms', the desire to suggest how exciting and intense it all is, to add a touch of romance. In reality, nothing could have been less like a storm than the neural firings of the mouse in the cage as they were presented to me, and no one, from a straightforward account of that electrical activity, could ever get any sense of the experience the mouse might have had of the world. For that it would be far better simply to watch the mouse moving around.

What I should have asked Monyer, then, was whether, in order to have this firing of the single neuron when the mouse had its left side to the wall of the cage at a particular point, it wasn't actually rather important to *have the wall there and the mouse's body crouched beside it*. The experience *was* the wall of the cage and the mouse's relation to it. As for the repeated and accelerated neuronal firing patterns while the mouse was at sleep or resting, they can only be a matter of speculation, since we don't know what experience the mouse was having, whether dreaming or daydreaming or neither, when these occurred. Perhaps these firings were delayed perceptions of some past situation. Or perhaps they have the function of forming, as we have suggested before, a kind of lock, ready to turn into recognition when the original experience is repeated. But perhaps not.

In any event, this gap between facts and storyline is endemic in the area of brain studies. One need only read a book like *Conversations on Consciousness*, or go to a website like closertotruth.com where world-renowned neuroscientists discuss, for example, the nature of memory, to observe the extraordinary difference between such experts discussing the nitty-gritty of their actual findings – full of competence and precision – and the same person speculating on what these findings mean, at which they become extremely vague, or defensively emphatic. How many neurons and synapses, the interviewer on closertotruth.com asks a dozen experts, would be necessary to establish a memory? And he uses the word engram: 'what is the minimal possible combination of synaptic connections that can constitute an engram', meaning, as Merriam-Webster has it, 'a hypothetical change in neural tissue postulated in order to account for persistence of memories'.

The replies he gets vary wildly, from a handful of neurons, to hundreds of thousands, from single cells dedicated to single memories, to neurons playing their tiny parts in infinite different circuits each constituted by millions of cells and each supporting different memories. But none of his interviewees remembered the qualification 'hypothetical' in the dictionary definition; engram is a word we have invented for a referent that may not exist. In this sense it is not unlike the word 'angel'. To believe in it is, for the moment, an act of faith.

Over lunch, Monyer talked engagingly about the world she moved in and the element of luck involved in winning prizes. And the following evening, when Eleonora and I met her in a bar of her choice for after-dinner drinks, she spoke interestingly of her family and the experience of emigration from Romania to Germany. When I floated the Spread Mind theory, the idea that experience was actually located outside the head, the experience being in fact one with the object experienced, and suggested that neuroscience might rig up some experiments to test this thesis, she was barely interested. But she did confess at one point that killing mice had always been a problem for her. She had never personally killed a mouse, she said, and often thought with sadness of all the mice that had been sacrificed to neuroscience.

Gad67EGFP Mice

We can't just see off neuroscience like this, can we, in a few pages? Its achievements are too many, its position in the public imagination, as the supreme arbiter on matters concerning the mind, too central. We must give it a little more space to lay out its wares. A year later I would come back to Heidelberg to take part in an evening when all the writers on our Science and Religion project got together to talk about the experts they had met and the conclusions they had come to. By that time I had already written half of this book. My ideas were clearer. I had read more widely. I had read another work by Christof Koch. I had read Giulio Tononi who believes that the secret of consciousness resides in the sheer complexity of 'integrated information', by which he means not only the extraordinary number of connections in the brain but also the particular nature of their organisation. I had read David Eagleman, who offers perhaps the best *reductio ad absurdum* of the neuroscience position on consciousness, but also gives a broad sense of the many and varied projects people in the discipline are undertaking. Reflecting on all this reading in the period before returning to Heidelberg, I made another appointment with Hannah Monyer, who I felt still had a great deal more to tell me and she again very generously sent me a number of papers she was publishing together with her team of six colleagues, one entitled 'Spatially Segregated Feedforward and Feedback Neurons Support Differential Odor Processing in the Lateral Entorhinal Cortex'. However discouraging that may sound, and certainly I was initially discouraged, you and I, dear reader, a little later in this chapter are going to spend a few pages getting our minds round this article. With great doggedness we are finally going to get a sense of what a typical neuroscience experiment really involves, the sort of results it produces, and the way those results are talked about in the ensuing publication. But before doing that it seems worth mentioning another, more general paper, published in the prestigious pages of *Nature* in April 2016. Entitled 'Neural Correlates of Consciousness: Progress and Problems', this article amounts to a summary of the present situation in the field of neuroscience, and in fact its four authors include both Koch and Tononi, two of the most celebrated names in the discipline.

The opening sentence of the paper's introductory paragraph takes as proven and undisputed the idea that perception is produced in your brain: 'There have been a number of advances', it claims, 'in the search for the neural correlates of consciousness – the minimum neural mechanisms sufficient for any one specific conscious percept.' This implies that we know that percepts are created by neurons and that neuroscience is simply trying to establish how many are required for a given perception.

The second sentence, however, tells us that previous hypotheses have proved wrong and have now been substituted with a new, spatially redistributed account of how perception is produced in the brain. 'In this Review, we describe recent findings showing that the anatomical neural correlates of consciousness are primarily localized to a posterior cortical hot zone that includes sensory areas, rather than to a fronto-parietal network involved in task monitoring and reporting.' In short, they've shifted the main action from the front of the head to the back. This is actually quite a change, suggesting how difficult it is to establish incontrovertible facts from the complex, often messy experiments neuroscience sets up. However, the vocabulary – 'task monitoring and reporting' – is clear and confident; the brain, it appears, is an administrative centre divided into sections working together on a variety of 'tasks'. It all sounds very business-like. You wouldn't be embarrassed to use these terms at the board meeting of Barclays Bank.

The third and last sentence of the abstract looks optimistically to the future while acknowledging some disappointment with what has actually been achieved: 'We also discuss some candidate neurophysiological markers of consciousness that have proved illusory, and measures of differentiation and integration of neural activity that offer more promising quantitative indices of consciousness.'

It's reassuring that even scientists are subject to error. They had ideas that research didn't back up. Fair enough. That's what science is about. It can't all be success. Even a brilliant idea can be quite wrong. Now, however, they have new 'measures of differentiation and integration of neural activity' – i.e. indications of how different neurons do different things and how these neurons are connected (though 'integration' is a slightly more loaded and mystery-yielding word than 'connection'). Here the abstract is looking forward to Tononi's claim that while consciousness may not be located in any specific neuronal activity, it might nevertheless emerge from the extraordinary complexity of the whole brain. It's hard not to note that this idea would seem to contradict the idea of the first sentence where it is suggested that one might isolate 'the minimum neural mechanisms sufficient for any one specific conscious percept'. How can that be the case if the complexity of the whole is necessary for consciousness to happen?

As the paper then sets out to describe the main developments that have occurred over recent years as well as the most active areas of current research, one soon senses the difficulty its authors are having establishing exactly what scientists believe and agree on. 'It has been known for a long time that being conscious requires the proper functioning of midline brain structures and that the particular contents of an experience are supported by the activity of neurons in parts of the cerebral cortex.' 'Midline' brain structures are simply those in the middle, near an imaginary line dividing the organ in two. But what does it mean that the contents of experience are 'supported' by the activity of neurons? This version could actually work well for either enactivism or the Spread Mind theory, neither of which deny the importance of the brain for allowing the world of experience to happen; just that proponents of these approaches don't believe the brain actually *produces* experience or that the experience is inside the head.

However, a little later this vague idea of 'support' is clarified. 'The content-specific NCC [neural correlates of consciousness]', we are told, 'are the neurons (or, more generally, neuronal mechanisms [interactions between neurons]), the activity of which determines a particular phenomenal distinction within an experience. For example, the NCC for experiencing the specific content of a face are the neurons that fire [produce electrical impulses], on a trial-by-trial manner, whenever a person observes, imagines or dreams a face, and are silent [don't produce electrical impulses] in other circumstances.'

You may wonder why I have taken the trouble to clarify easy words like 'fire' and 'are silent'. The answer is that it seems to me these words suggest an independent agency, almost a personality, on the part of the individual neurons or groups thereof, a sense of purpose, as they fire or keep mum. Metaphors are always tricky and one would think that when reporting scientific experiments this kind of contamination could be avoided. Simply, there are or are not electrical impulses. In general, the authors are suggesting that it is the activity of the neurons that determines the distinction between the faces, whereas one might reasonably wonder if the neurons are not simply reacting to faces that are in themselves distinct. Or rather, the face is required as much as the neurons. But no, the passage continues: 'When the content-specific NCC neurons in this example [face recognition] are activated artificially – for example, by transcranial magnetic stimulation (TMS), electrical stimulation or optogenetic stimulation – the participant should see a face even if none is present, whereas if their activity is blocked, the participant should not be able to see a face even if one is present.'

This is an important claim, then, that supports the notion that experience is created entirely in the brain. You tickle the neurons in the face-recognition area and people see faces willy-nilly; you inhibit them and people cannot see faces at all. Critics have pointed out that the faces that participants in experiments of this kind are induced to see by having their neurons electrically stimulated are faces they already know, or made up with features they are familiar with, hence dependent on past experience; the face had to be there, physically, once. They never see features they have never seen before. So a theory like the Spread Mind would put such a phenomenon in the same class as dreams and claim that our perceptive apparatus was still causally affected by the original object. The paper's use of 'should' and again 'should not' in talking about the reactions to stimulation and paralysis of this area – they *should* see a face when . . . they *should not* see a face when . . . – suggests that actually participants do not always respond by seeing or not seeing faces, and hence that there is something a little perverse in their not reacting as the scientists' rule would suggest. In general, the logic here is that scientists *should* be able to re-create, or recall, more or less every experience by stimulating our brains in certain ways. However, there are not many accounts of this actually occurring.

In a following paragraph, to complicate matters, the authors admit that much of the neural activity that appears to correlate with specific and simultaneous experiences can be seen to be going on *both before and after* those experiences. They presume this can be explained by activities 'such as selective attention, expectation, self-monitoring, unconscious stimulus processing, task planning and reporting, rather than the experience itself'. This is fair enough, but it is no more than a reasonable supposition, a narrative they are imposing on the activity they have observed. Again one notices that words like 'planning' and 'reporting' suggest that we are talking about a number of independent entities that discuss matters with each other; apparently this is the only way we can talk about complex organisation. Meantime the expression 'unconscious stimulus processing' is rather mysterious. It is common knowledge that much of what goes on in and around our bodies is not at the centre of conscious attention and perhaps not even in the zone of attention at all, hence unconscious, but what does it mean to say it is unconsciously 'processed'? When a computer chip processes the input we feed through it, it does so in line with software we have written in ways we can predict. It's not clear whether this scenario is analogous with what is happening among our neurons.

Another part of this prestigious paper considers the neural correlates of vision in general, that is, the neurons that are active simultaneously with visual experience (which of course goes on almost all our wakeful hours and hence is difficult to isolate in a single moment). 'Whether the primary visual cortex (V1)', it begins, 'contributes to visual consciousness directly or whether it has only an indirect role – much like the retina's role in visual consciousness – is the subject of ongoing debate.' In short, we don't know. However, we do know of cases where someone whose V1 (an area at the very back of the brain) has been damaged reports not seeing 'an item' but nevertheless performs 'above chance on forced-choice tasks'.

What does this mean? That although someone with a damaged V1 may *say* they can't see this or that object, nevertheless, if you create an experiment in which they are invited to push a button to indicate whether the item is there or not – a forced-choice task – their score will be higher than if they were blindfold (though much lower than that of the ordinarily sighted person). The paper reflects: 'Their subjective blindness could be a result of the insufficient feedforward activation of higher visual areas [from V1], or to lack of feedback to V1, in which case V1 would be necessary for conscious vision.'

'Subjective' here would appear to mean that these people *believe* they are blind when actually they are not, or only partially; this seems an extraordinary way to approach this unhappy condition, since, in the internalist view of consciousness, all perceptual experience is necessarily subjective in that it is a personalised representation of reality, not reality itself. How could it ever be objective? Anyhow, if you don't see the object, you don't see it, and you know that, even if something is still going on to allow you to score above random response in a 'forced-choice task'.

The rest of the sentence speculates that this form of half (or less than half) vision may be due to the V1 area failing to send on impulses to other areas or failing to receive impulses from them. This seems like intelligent speculation. Very likely they are right that the damage has interrupted connectivity between different parts of the brain involved in different activities. However, the whole question of how the brain is involved in the experience of seeing remains a huge conundrum. That is, after about fifty years of modern neuroscience, despite an enormous amount of data having been gathered, very little has actually been clinched. The paper concludes:

> In short, no single brain area seems to be necessary for being conscious, but a few areas, especially in the posterior cortical hot zone, are good candidates for both full [general] and content-specific neural correlates of consciousness . . . The sheer number of causal interactions in the brain, together with the fleeting nature of many experiences, pose challenges to even sophisticated experimental approaches to the NCC. The NCC by themselves can provide little information about consciousness in patients with severe brain damage, infants, fetuses, non-human species or intelligent machines [that is, tracking neuronal activity doesn't tell you much about what is being experienced – quite how intelligent machines get into the list I'm not sure]. Further progress in this field will require, in addition to empirical work, testable theories that address in a principled manner what consciousness is and what is required of its physical substrate.

The last sentence here is simply extraordinary. These celebrated scientists first admit that they lack 'testable theories' of what consciousness is, but immediately go on to posit that there is consciousness on the one hand and a 'physical substrate', or underlying material basis, of consciousness on the other, suggesting that consciousness is *not* physical. This in itself amounts to a theory of consciousness, and what's more one that seems inescapably dualist. You have the physical basis on one side and the mental experience, whatever that may be, on the other. I don't want to labour this, but it does seem important to establish once and for all how *muddled* a group of scientists can be while nevertheless occupying centre stage in the collective imagination as far as ideas about consciousness are concerned.

With regard to occupying the centre stage, in October and November of 2015 the American Public Broadcasting Service aired a TV series written by the neuroscientist David Eagleman entitled *The Brain, the Story of You*. The programme was shown in the UK in February 2016 and then presented in book form later that year. While Koch and company in *Nature* are properly cautious as to what has been achieved, Eagleman, who has been trying for many years to popularise the ideas behind the discipline, addresses the general public as though he possessed the truth. Before returning to the more serious Hannah Monyer and her mice, let me string together a few quotations from Eagleman's book.

> Your brain serves up a narrative – and each of us believes whatever narrative it tells. Whether you're falling for a visual illusion, or believing the dream you happen to be trapped in, or experiencing letters in color, or accepting a delusion as true during an episode of schizophrenia, we each accept our realities however

our brains script them. Despite the feeling that we're directly experiencing the world out there, our reality is ultimately built in the dark, in a foreign language of electrochemical signals. The activity churning across vast neural networks gets turned into your story of this, your private experience of the world: the feeling of this book in your hands, the light in the room, the smell of roses, the sound of others speaking.

In short, your experience has little to do with reality. You are fooled, but Eagleman, despite the fact that he also has a brain (one supposes), is not. He goes on:

What if I told you that the world around you, with its rich colors, textures, sounds, and scents is an illusion, a show put on for you by your brain? If you could perceive reality as it really is, you would be shocked by its colorless, odorless, tasteless silence. Outside your brain, there is just energy and matter. Over millions of years of evolution the human brain has become adept at turning this energy and matter into a rich sensory experience of being in the world.

Following Galileo, whom he contrives never to mention, Eagleman tells us that colour, smell, sound and taste are inventions of the brain. They don't exist. It's a show. Consequently, 'who you are at any given moment depends on the detailed rhythms of your neuronal firing'. And on the following page, since Eagleman doesn't disdain emphasis and repetition: 'who you are depends on what your neurons are up to, moment by moment'. Or again a little later: 'Your interpretation of physical objects has everything to do with the historical trajectory of your brain – and little to do with the objects themselves.' Then once more: 'Our perception of reality has less to do with what's happening out there, and more to do with what's happening inside our brain.' And finally, just in case you still haven't grasped the idea: 'All of your sensory experiences are taking place in storms of activity within the computational material of your brain.' Meantime of course your brain 'has no access to the world outside. Sealed within the dark, silent chamber of your skull, your brain has never directly experienced the external world, and it never will.'

How dramatic!

Readers may feel a little confused hearing one moment that the brain tells them stories, suggesting a gap between brain and self, and next that their identity is a question of neuronal activity, which would drastically close the gap. But perhaps it doesn't matter, since Eagleman's main concern is simply to establish the primacy of the brain and hence of neuroscience. Whatever may or may not be happening out there, he insists time and again, the real location of the drama (even when you have no awareness of drama at all) is inside your head. Experience is 'an electrochemical rendition in a dark theater'. Decisions – like whether to choose lemon- or mint-flavoured ice cream – are pitched battles between 'warring networks of neurons'. 'They fight it out until one triumphs in the winner-take-all competition. The winning network defines what you do next.' In fact: 'Thousands of choices bear down upon shoppers, with the end result that we spend hundreds of hours of our lives standing in the aisles, trying to make our neural networks commit to one decision over another.'

How befuddling is this? We don't choose between the products we see on the shelves, or believe we see, the products, that is, that very soon we will be eating (Eagleman does not deny that we really eat them). We try to get neural networks to commit. But wasn't our identity supposed to depend on the neural networks, indeed be constituted by them? None of this makes any sense at all. Yet it is the story that is served up to the public by prestigious organisations like the PBS and the BBC.

But back to Heidelberg. Monyer had sent me her papers, I read them carefully, checking all the new vocabulary and studying all the various techniques and processes she referred to and then we made an appointment to talk in a coffee shop on a cold Saturday afternoon. She arrived, late, on a bike, full of bounce and energy. We grabbed the two last places at a table for a dozen and began to speak in loud voices about mice and their sense of smell. An hour later we were alone . . .

Here then, as well as I can describe it, is an account of the experiment. Again, what I am inviting the reader to do is to step behind the glitz and sleight of hand we have to put up with when these things are presented to us by David Eagleman on TV or in the popularising, wow-isn't-this-amazing literature, and try to assess what has really been achieved, what is really learned and known in the process of an experiment like this.

First we're talking about a place, the lateral entorhinal cortex, or cortexes, since there is one at each side of the head. To save space we'll sometimes call it the LEC. The entorhinal cortex (where 'rhinal' means having to do with the nose, and 'ento' indicates inside) is close to the hippocampus at the core of the brain. We've mentioned the hippocampus and its presumed role in spatial awareness and memory before. In the image below, taken from Wikipedia, the entorhinal cortex corresponds to the numbers 28 and 34:

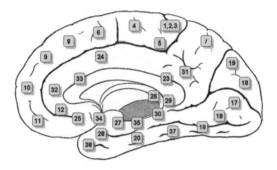

This obviously is a human brain. The diagram overleaf (top) gives its position as EC in various brains.

The LEC's position in rats and mice is particularly fortunate for laboratory technicians in that it is right there at the side of the skull and easy to access. Below the brain diagram is a drawing showing how the area was accessed for the experiment we will be looking at. It's hard not to feel a certain sympathy for the mouse seeing it trapped between all these threatening lines and words.

The entorhinal cortex, we should know, is divided into various parts or layers, the lateral entorhinal cortex being the part immediately inside the skull. All the various

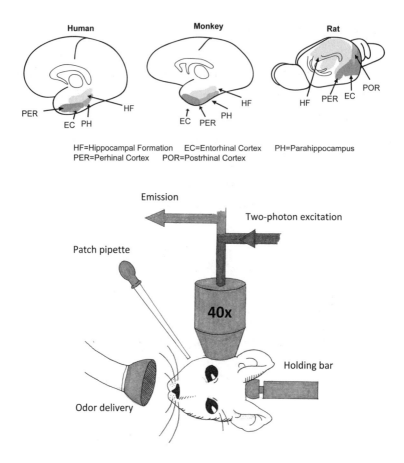

HF=Hippocampal Formation EC=Entorhinal Cortex PH=Parahippocampus
PER=Perhinal Cortex POR=Postrhinal Cortex

encyclopaedia sites carrying the textbook wisdom on such matters are confident that the entorhinal cortex as a whole 'functions as a hub in a widespread network for memory and navigation' constituting 'the main interface between the hippocampus and neocortex'. The layer in the middle of the area, the medial entorhinal cortex, has neurons which respond to space and movement in space. The neurons in the lateral entorhinal cortex, however, respond mostly to smells. So the introductory abstract of our paper begins with the announcement: 'The lateral entorhinal cortex (LEC) computes and transfers olfactory information from the olfactory bulb to the hippocampus.'

Needless to say this single sentence implies volumes about how its authors suppose the brain 'works'. What on earth, I ask Monyer, does the word 'compute' mean. And once again we have that bugbear 'information', as if we were referring to something substantial, or some identifiable content, that gets processed, repackaged and sent along, as if to some ultimate place of consumption and reflection, which again seems like a hierarchical model of management and very likely not at all analogous with what is happening in the brain. I ask Monyer why neuroscientists are so attached to these metaphors; you would never get a biologist talking about cells in the liver 'computing' things. She laughs and sighs and agrees that these words are misleading, but *only for the layman*. 'Neuroscientists aren't misled because we know what is meant,' she says.

'So why don't you just say what you know is meant?'

She has no answer to this. My suspicion is that whatever they may know about what actually happens when they use the word 'compute', nevertheless they are living and moving inside the dominant internalist model which assumes the brain is some kind of supercomputer; hence all their experiments are structured around this idea. The next sentence of our abstract confirms it. 'We established LEC connectivity to upstream and downstream brain regions to understand how the LEC processes olfactory information.'

But rather than take issue once again with the notion of

'information processing', let's look at what exactly is being done in this experiment. First the team does a lot of dissection, colour staining and brain slicing to establish exactly what kind of neurons are present in the LEC of mice, in what numbers and how distributed. This leads to their focusing on three dominant kinds of neurons: RE+ (reelin plus) neurons, CB+ (calbindin plus) neurons and GABAergic neurons.

Reelin and calbindin, we discover, are proteins, and RE+ and CB+ neurons 'express' or produce these proteins. Reelin is known to favour the plasticity of the synapses; calbindin, as its name suggests, binds, or captures, calcium. Fortunately, these two groups of neurons, which form about 80% of the overall population, are not mixed promiscuously together in our mouse LEC, but form quite distinct groups in slightly different areas, layers even. Both types are what are known as excitatory rather than inhibitory neurons, that is, the impulses they send out encourage further impulses to be sent on from the neurons they connect with. The GABAergic neurons on the other hand are inhibitory neurons, and these are distributed fairly evenly in amongst the other two groups. The reason for this distribution begins to make sense when you understand that excitatory neurons tend to connect up with neurons some distance away, often in different parts of the brain, while inhibitory neurons, which tend to block or dampen down activity in the neurons they connect with, tend to connect locally. So what you have is about a fifth of your neurons evenly distributed and blocking, or not blocking, the excitatory activity of the larger groups of neurons which are essentially separate from each other and sending impulses far afield. This kind of arrangement is fairly typical, it seems, across the brain and suggests why people like Eagleman have conjured up metaphors like 'warring' neurons, though of course there's no reason at all to start thinking in terms of conflict. What we have is a complex, self-modulating organisation. It could hardly be otherwise.

So the next job is to find where exactly the excitatory RE+ and CB+ neurons are connecting *to*. This is done by injecting different tracer substances (fluorogold and cholera toxin) in likely areas of the living brain and seeing if the traces then arrive in our LEC, and if so, in which neurons they appear. Obviously this means more mice and more brain slicing, and long hours of work as the team examines neuron after neuron *individually* (remember that around a million have the same volume as a grain of rice), checking to see whether each neuron has taken up the tracer. For each result the published paper gives the exact number of neurons counted: 1,097 fluorogold positive neurons were counted in the area of the RE+ neurons; 1,107 cholera toxin positive neurons were counted in the area of the CB+ neurons. And so on. Eventually it emerges that RE+ neurons connect 'downstream' (further away from the original sensory impulse coming from the nose and olfactory bulb) to the hippocampus (in particular the dentate gyrus area of the hippocampus); so they are *feedforward* neurons; CB+ neurons connect back 'upstream' towards the olfactory bulb where the original impulses came from, so they are *feedback* neurons. However, it's worth noting here, just to get a sense of the mad complexity of the brain, that although the distinctions between the neurons are strong, they're not absolute. A small fraction of the fluorogold positive neurons were GABAergic, suggesting that some of those inhibitory neurons are connecting further away than was suspected, and so on. In this phase the team also discovers, with more injections and more slicing, about half of the CB+ neurons also connecting to the other lateral entorhinal cortex on the other side of the brain. In short, there is a lot of back and forth going on.

Having established what neurons are where and where they hook up to, the team is now ready to prepare for the main part of the experiment which will involve anaesthetising the mice and subjecting them to six different chemical smells – ethyl butyrate, amyl acetate, eugenol, cineole, hexanal and benzaldehyde – while measuring the response of the neurons, in vivo in real time. How on earth can they do this? Well, the simple part is the insertion of a window in the mouse's skull. Here is the paper's description of how this is done:

Mice were anesthetized with a mixture of ketamine and xylazine (120 mg/kg, intraperitoneal injection). Additional doses were administered through the course of the surgery if necessary. Dexamethasone (0.02 ml at 3 mg/ml PBS) was intramuscularly injected into the quadriceps to reduce cortical stress response during the surgery and prevent cerebral edema [swelling]. Lidocaine was administered locally to the skull as additional anesthetic for pain reduction. Animals were mounted in a stereotactic apparatus [this allows machine-guided, super-precise surgery] and placed on a heat pad to keep the body temperature constant. A small cut of approximately 1.5 cm was made in the skin and the underlying skull revealed. The skull was cleaned of periosteum [a membrane that covers bone], and muscles located laterally at the right side of the skull were removed. Four miniature screws were screwed into the skull and fixed with dental cement. Orthogonal [at right angles] to each other, two cap nuts were cemented on top of the screws as anchors for holding bars, which allowed tilting the mouse by 90° to gain direct access to the LEC. A head plate was secured to the skull using dental cement and a small craniotomy [removal of part of skull] was made. Regular rinsing with normal rat Ringer solution prevented the brain from overheating due to drilling. The dura mater [a thick membrane protecting the brain] was removed leaving the underlying brain surface intact. Agarose [a gel] was applied to the imaging window and a small round cover glass (thickness 100–150 µm) was placed on top. During the time of recording neuronal activity, urethane was used for keeping the mouse anesthetized.

One can see why these experiments are expensive. This delicate surgery was carried out on at least thirty mice (the paper is never clear as to the overall number of mice used in the experiment). Of these some were what is called 'wild-type mice', that is mice that haven't been in any way genetically modified, while many were Gad67EGFP mice. Quite a name. To go into detail as to what this means would require a chapter and more all to itself, but essentially they are mice genetically modified in such a way that their GABAergic neurons 'selectively express enhanced green fluorescent protein (EGFP)'. So we're talking about a technique for labelling the neurons and hence understanding which are doing what. In addition to using some genetically modified mice, the team also injects a virus which will induce other neurons to produce GCaMP6, another fluorescent indicator. Again the details of the chemistry and biology involved are beyond my explanation. Suffice it to say that the technology has been around since 2004 and that in this case it involves different viruses and tracers being injected into the different kinds of mice. Leaving aside another long description of the surgical process involved in delivering these viruses locally, we can say that essentially a micropipette is lowered 2.3 mm below the surface of the brain, into the LEC obviously, then retracted to 1.6 mm 'while continuously injecting the virus (300 nl) for a period of 5 min'. An 'nl' is a nanolitre, a unit of volume equivalent to 10^{-9} litres. It takes quite an effort even to imagine the precision and delicacy of what is involved here.

After the injections, seven days must pass for the viruses to do their work. One hopes the mice enjoy this week, since it will be their last. At this point, with the rodent brain chemically prepared in such a way that any neuronal activity in the LEC will show up as fluorescence, the real experiment can at last begin. First the mouse is anaesthetised then 'tilted by 90° and fixed using a holding bar screwed to the cap nut on the top of the skull'. A 'two photon' microscope is then set up over the window in the skull; again, one has to read many pages to get some grasp of what a two-photon microscope involves, but basically we have an apparatus that can both shine lasers into the brain and simultaneously photograph and record the fluorescence or otherwise of the neurons in the surface area of the LEC that has been exposed, the recordings being of course synchronised with the process of delivering the odours to the mouse's nose, this so that we can measure how long it is after the release of the odour before the mouse, or rather its neurons, respond (the mouse qua mouse being incapable of any response), and so on. Needless to say, not all the mice give good results. Bleeding can be a problem. So certain mice 'were excluded from further analysis'.

The wonderful two-photon microscope, however, is not capable of accurately measuring the fluorescence of slightly deeper tissue in the brain, and while the RE+ neurons are right on the surface, the CB+ neurons are in the layer beneath them, so for these neurons the team apply pipettes directly to neurons under the surface, using the patch-clamp technique described in the previous chapter. There is now a lot of expensive clutter around our anaesthetised mouse.

How are the odours delivered? 'A custom-built flow-dilution olfactometer mixed saturated odor vapour of pure odors with filtered air. The carrier stream of clean air was set to a flow rate of 2 liters per minute, which was mixed with odorized air of 0.2 liters per minute for a final concentration of 10%.' All this as a general rule, but other concentrations were tried too, as well as diluting certain odours in mineral oil. It only makes sense, once one has made the investment in the overall project, to try various permutations. This is not the kind of study whose limits are fixed at the start; techniques can be changed on the go in response to the kind of results that emerge, or obstacles that arise. In fact I fear I have told less than half the story and that each of the steps described so far involves infinite adjustments and refinements to make sure that we get the best possible measurement of what a mouse's LEC neurons do when you hit the creature with strong smells.

Enough digression. Each odour is delivered ('by a tubing system and solenoid valves under computer control') for four seconds, the tube's aperture being positioned 'less than one centimeter from the animal's nostrils'. The team calculates that with this method the delay in 'odor onset', that is the time between the release of the smell and its arrival at the mouse's nose is 330 milliseconds, something measured using a photoionisation detector (readers wishing to know how this precious machine works can check it out online), and of course this is taken into account in the synchronisation of the delivery apparatus with the recordings made by the two-photon microscope.

Each four-second blast of smell (foul for the most part) is followed by a period of twenty-six seconds of clean air, 'to avoid sensory adaptation', that is, so that the mouse starts from zero, more or less, for each smell. In fact only three smells are delivered in each experimental bout (meaning a period of ninety seconds: $4 + 26 \times 3$) with rest periods in between before the other three smells are delivered. Each bout is repeated fifteen times, which means thirty bouts in all. At this point (the mouse now having been hit with smells ninety times), the two-photon microscope is moved slightly to look at another area in the brain window and the bouts are repeated another fifteen times, for each group of three smells (so, another thirty bouts), the order of smell delivery being randomised for each bout, again 'to avoid sensory adaptation'. For some mice the microscope will be shifted just once, for others twice, for others three times, so that a total of four brain areas can be investigated during the experiment. In this last case, then, the mouse is tested with a total of 120 bouts, encountering strong smells 360 times.

Just describing this, I can't help feeling how surreal, how mad it is all becoming, how obsessive one has to be to establish anything at all in this world of minute, complex, fleeting activity, made up of billions of neurons with their trillions of connections. Sitting in the pleasant Heidelberg coffee house with Hannah Monyer, eating an excellent *Apfelstrudel*, I did ask her whether an anaesthetised mouse could possibly give the same response that a mouse scurrying around free, in a darkened vicarage boiler room for example, would give, and she said obviously not. Mightn't it perhaps react differently, I asked, even under anaesthetic, if the smells were smells that meant something: cat smells, sex smells, food smells? It very well might, she laughed. I also wondered whether the same neurons responding here to the smells, might not, under ordinary circumstances, also be responding to other 'environmental factors' (one starts to talk like this after reading a few scientific papers) and she said yes, very likely they would, and that was precisely why they had to be anaesthetised, because if they weren't it would be

extremely difficult to be sure that the neurons were responding directly to the smells, not to mention the impossibility
of looking directly in their brains. There were certain experiments that could be done with the mice running free (in
cages) – for example the experiment I had observed in the
lab the previous year relating to the position of a mouse in
space – but not many. There was even, she told me, laughing, a rather crazy German neuroscientist who loaded his
mice with recording gadgets and set them free in a field, but
the problem was that it became almost impossible under
these circumstances to sort out all the data that you got,
since the neuronal activity might correlate with any number
of things. A mouse's life is a busy one.

Even in a highly controlled experiment like this, sorting
out data is extremely complex. Any notion that on seeing
the photos included in the published paper you are simply
seeing a magnification of what you might see looking into
the mouse's brain would be quite naive. In the pages-long
explanation of how the various kinds of data were treated, a
couple of sentences read as follows:

> Calcium imaging movies were warped to high-
> resolution reference images by custom-written routines
> based on nonlinear image registration. For each ROI
> [region of interest], the average fluorescence intensity
> was extracted and converted into relative change in
> calcium according to

$$\frac{\Delta F}{F} = \frac{F - F_o}{F_o}$$

[This essentially means that the difference between initial fluorescence intensity and intensity after odour stimulation is divided by the value of the original intensity.] Baseline fluorescence Fo was calculated as the tenth percentile of background subtracted fluorescence for each ROI. Mean $\Delta F/F$ level during the 4 second pre-stimulus period was subtracted from each individual odor response. Odor-evoked receptive fields were computed by averaging calcium responses to all presentations of a specific odor and sorted from 'best' to 'worst' based on the mean $\Delta F/F$ value during a time window two to three seconds after stimulus onset (where typically the early response peak occurred).

In short, an awful lot of effort is required to make sense of the results. This is because each neuron responds to each smell in different ways and these responses are themselves constantly changing over a period of some seconds. For example, about half the neurons examined, after following the regular pattern of a peak of reaction at two to three seconds after the onset of the smell, then a decline, 'displayed an additional late fluorescence increase following stimulus offset', i.e., after four seconds. They reacted, that is, to the smell's disappearing *in the same way*, but with less intensity, as they had reacted to its appearing. Why? And why only half the neurons?

To answer this and a hundred other questions, after the experiment was concluded, the mouse's brain was extracted and sliced, following the procedure described in the previous chapter, and the various neurons involved investigated, where possible individually, to see if there were morphological (shape) differences or electrophysiological differences (to do with chemical make-up and electrical charge) between neurons responding differently. But let me now try to give an overview of the general results so far as I have understood them; and Hannah Monyer, over strudel, assured me that I had in fact, to my surprise, understood them pretty well.

So: overall around 80% of neurons reacted to at least one odour. Of those reacting, about 77% did so with increases in fluorescence, but 37% reacted in some circumstances with a decrease (meaning some neurons showed increased fluorescence with some smells and decreased fluorescence with others). Gratifyingly, the three kinds of neurons under examination reacted rather differently. The RE+ neurons – which, remember, connect deeper into the brain – reacted the most strongly and selectively; that is, they sent stronger electrical impulses, or patterns of impulses (for a couple of seconds) and reacted more definitely to 'preferred' odours, and hardly at all to others. The CB+ neurons, which connect back to the area where the smell is coming from and/or across to the LEC on the other side of the head, were a little less selective and receptive, while the GABAergic neurons responded at a much lower level and reacted in much the same way to all the smells, as if it was the fact of smell in general rather than the particularity of the smell that affected them.

More brain slices were examined to see what differences there were between the RE+ and CB+ neurons responding to particular smells, and it was established that their axons and dendrites seemed to head to slightly different places. There was much repeating of the experiment and the analyses to try to nail this down. In particular it was established that the CB+ neurons were connecting to, and hence exciting, neurons in the olfactory bulb area that were actually *inhibitory* in their function. So the effect of these feedback neurons seems to be to correct or at least change the kind of response occurring back at the odour entry point, which then affects, of course, the impulses received in the LEC. Hence we have a loop. However, it was also established that the CB+ neurons were already sending impulses to the olfactory bulb before the onset of any smell.

As for the GABAergic neurons, two kinds were identified, one of which had the effect of inhibiting the other, which thus no longer inhibited, or less effectively inhibited the RE+ and CB+ neurons they were connecting with. Again, an intricate series of checks and balances seems to be in operation and again much examination of brain slices was undertaken to establish whether these cells had different morphologies and electrophysiological features. Curiously, it turned out that there was no evident correlation between a neuron's morphology and its electrophysiological characteristics.

But to return to the general results. Since the delivery of each odour resulted in a particular pattern of response from each kind of neuron, a computer was programmed to analyse the separate responses – the number of electrical impulses, the time lapse after onset, the peak, the nature of the decline, the possible upward blip after offset – then fed data from new tests and asked to identify which smell was which on the basis of the neuronal response. And this it managed to do, particularly for the RE+ neurons and particularly when considered as a group rather than singly. That, for sure, was a remarkable success indicating that neuronal response does correlate in patterned form, particularly when considered as an entire population rather than as separate individuals, to particular smells.

Numerous small coloured photos with patterned dots, intricate colour-coded bar charts and attractively naive hand drawings of neuron shapes, with their long straggly root-like axons, followed by photos with the same drawings superimposed on them in miniature to show where these particular neurons are, or were, located, in the LEC, accompany the published paper describing the experiment and allow for hours of study as you try to figure out exactly what they represent and by what means they have been achieved. At the conclusion of the main paper (and before some fifteen pages of detailed accounts of the techniques and procedures used), the team reached the following conclusion (the italics are my own):

> The *decoding performances* of both RE+ and CB+ populations are better than those of GABAergic neurons and enable *the extraction of odor identity* from the population activity. RE+ populations in particular display the *highest information content*, possibly because they comprise a mix of highly selective neurons responding to distinct odors. Thus, we suggest that *upstream and downstream regions* of the LEC in the olfactory system are supplied with odor information via

a *population code* rather than by single cell activity in analogy to what has been proposed for odor encoding in the PIR [piriform cortex, the part of the brain in immediate contact with the olfactory bulb].

In summary, two distinct odor-responsive excitatory cell types are spatially segregated in LEC LII. Their long-range connectivity and *differential odor representation* indicate that they subserve different functions. RE+ neurons *transmit odor information* to the hippocampus and *might serve the formation of associative memories*, whereas input of CB+ neurons to the OB [olfactory bulb] *might set the threshold to responsiveness* to odors, thus leading to adaptation of sensory responses.

What can one say? The team has done an astonishing job teasing out all kinds of data from the tiny brains of their rodent specimens using batteries of extremely sophisticated equipment and drawing on a huge accumulation of knowledge and know-how in physics, chemistry and biology. It really is amazing that mankind has reached the point that it can do these things. There is a wonderful, wonderfully implacable intelligence about how an experiment like this is set up and how the team responds to the results they are getting to refine their goals and methods. But 'decoding performance'? 'Extraction of odor identity'? 'Information content', 'population code', 'differential odor representation'? How do these formulas fit in with the concrete data gathered?

It is true that a computer, having been shown how the neurons react to the six different odours, and told which reaction relates to which odour, can then identify further reactions as being specific to this or that odour. This shows that there is consistency in the neuronal response to the different smells. But it hardly shows they are 'coded'. The computer doesn't analyse the data in relation to what it knows about the chemical make-up of the odours. There is no connection that anyone can determine between the different neuronal firing patterns in response to the smells and any chemistry we know of. The computer can't predict what smell – beyond the six recorded – the neurons were reacting to, by extrapolating a code from the way they react to these six. What can it mean, then, to say that the RE+ neurons enable the 'extraction of odor identity'. Who is 'extracting' it? Who is labelling it with an identity? We are perilously close to a homunculus model of experience here.

And what can it mean to talk about 'information content' and 'transmitting odor information'? There is nothing in this experiment to suggest that a neuron receives a given piece of data and passes it on, perhaps after altering it a little ('processing it' would of course be the preferred vocabulary). The anaesthetised mouse is hit with powerful smells. The olfactory bulb connects to the LEC. The neurons in the LEC respond in slightly different ways to each smell, firing off impulses backward and onward. That is what we know.

What does it mean, in these circumstances, to talk of an odour *representation*? Even conceptually, this is hard for us. We are used to the idea of visual and auditory representation because we have learned how to make drawings and photographs of objects and to record sounds and replay them when we choose, separate from the original sound. But what is the *representation* of a smell? Why not say the experience of the smell? I put it to Hannah Monyer – and she had ordered us second coffees by this time, coffee that had a wonderful aroma – that the conclusions were saturated in a particular vision of how consciousness works, a vision which thinks of the brain as forever encoding and decoding but rarely actually *experiencing* the world, a vision which always leaves us with a chasm between the technical reporting of electronic impulses and chemical exchanges and the nature of experiences. Upon which, and I must say that she was more than game for the discussion today, she told me that she had decided way back, right at the beginning of her career in fact, that there was simply no point in working on consciousness, because she couldn't see how neuroscientific experiments could ever yield results in that field. She was interested in memory.

'But memory is part of consciousness!' I objected. 'The awareness that you have smelt this smell before is a conscious awareness.' Though I then immediately had to concede that the body could react differently a second time to something – as Sabina Pauen's children react differently the second time they see a photo, without knowing they have reacted differently. We can 'remember' perhaps, without being conscious of remembering. 'I can see,' I said, 'how on the basis of these results, you could hypothesise that the feedback of "CB+ neurons to the OB might set the threshold to responsiveness to odors", that makes sense, but I can't see what there is in this experiment that prompts the claim that RE+ neurons "might serve the formation of associative memories".'

Monyer smiled and said that it was common knowledge associative memories were formed in the hippocampus, so neurons sending information to the hippocampus could be seen as perhaps serving to form such memories. She was trying to trace all the activity necessary to the formation of a memory. All the same, she agreed, there were still a lot of 'perhaps's and 'might have's. 'The fact is, though,' she repeated, 'we neuroscientists know the reality we are describing when we use this vocabulary.'

So now at last I put it to her that there was absolutely nothing in the data the experiment produced that was not compatible with an entirely different model of consciousness, where the neurons, rather than *representing* the smell in the head somewhere, whatever that might mean, were part of a perceptive apparatus that permitted the smell to happen, as experience, at the point where it was located outside the body. This would save the search for a point in the brain where information is finally decoded and representations are finally rehearsed for the benefit of some unfathomable and invisible self. It could even lead to a whole new approach to experiments whereby one thinks of neuronal behaviour not as encoding data to produce some final effect in the head, but as providing the base, the platform, for the world outside to assume the form it does. And I gave her Riccardo's new analogy of the dam and the lake. A dam is not a lake. They are separate. Only you need the dam for the lake to happen. In the same way you need the body, the brain, the eyes, ears, nose, neurons and so on for experience to happen. But experience remains outside, as the lake remains a different thing from the dam, relying on it but not identical with it.

Hannah Monyer thought about this as we drank our second coffee. She was taking the Spread Mind theory a little more seriously this time, perhaps because she was impressed that a layman like myself had made the effort to get his mind round one of her experiments. This was very rare, she had acknowledged earlier in the conversation; to an extent it caught her by surprise. Perhaps she should think a little more about these things, she said. On the other hand, the area of her expertise was finding experimental methods to probe the sophisticated workings of the brain, regardless of any model of consciousness. That was her vocation.

It was time to step back from the debate. We talked about our personal lives. She was deeply upset, she said, and a little disoriented by the loss of a colleague who had been close to her throughout her working life. They had inspired each other. Their research had been a joint endeavour and it was hard now that he was gone. But she had recently been given a new appointment running a research lab in Berlin and this extra work and the travel to Berlin would hopefully get her back in the saddle.

I was struck again by what a vital person she was, what an intensely lively mind she had. But at the same time I could not help thinking of the likewise intense, if rather more primitive lives of mice, of all the mice injected and mutilated and anaesthetised in laboratories all over the world. And of the way the more we cut them up the more we realise how astonishingly complex their brains and nervous systems are and how little we know about the nature of their experience. And though I have never been an animalist, or a supporter of animal rights – having always believed that if research using animals is carried out in such a way that they do not suffer and if such research saves and improves human lives, then it is legitimate – nevertheless, walking home now from the coffee shop, after agreeing with Hannah that we would stay in touch, I began to wonder if it really was *right* to cut up all those mice to learn what in the end any number of experiments have already told us, that the brain is full of neurons exchanging chemicals with each other and sending impulses back and forth. What can it mean to say 'in such a way that the animals do not suffer'. Who can know? And who is to set a value on the happiness or otherwise of a mouse?

There is a formula of words I learned long ago, at the first meditation retreat I went to, I don't know if you would call it a prayer, but it is Buddhist-inspired, a formula I repeat at the end of my early morning meditation sessions, that goes like this: 'May all creatures be happy, may all creatures be filled with joy and joy for the joy of others, may all creatures be free from all attachment, may all creatures be free.'

It is a little embarrassing to admit that I repeat this formula every day. How could it ever happen that all creatures could be happy and free, free from being attached to pleasures and pains in a way that will bring suffering, *dukkha*? It couldn't. Yet I do say it. And it *does* make sense to say these words after an hour's meditation. And in the days after studying that paper on 'Spatially Segregated Feedforward and Feedback Neurons Support Differential Odor Processing in the Lateral Entorhinal Cortex', I could not help thinking of all the mice in cages in laboratories waiting to have electrodes sunk into their brains and windows inserted in their skulls and could not help wondering what it means when there is happiness or there is pain, freedom or entrapment, even for the most insignificant of creatures – but who decides who is significant? – and wondering whether there mightn't be a different way forward for the human race, which didn't involve this slaughter, and how one could ever be part of a movement towards such change. Could, perhaps, the idea that our experience is not locked in our heads, but is really out there, the idea that we in a real way *are the world*, the world we know, change any attitudes, make us more respectful of that world, make us wonder if we don't do damage to ourselves when we do damage to others, to wild-type mice and even Gad67EGFP mice? If that is the case, then very likely it is one more reason why this idea will be resisted.

Schläft ein Lied in allen Dingen

There are days when one is seriously tempted to give up. There is always someone who has said it before you, and better than you, or whose ideas, discovered late in the day, threaten to undermine the picture you've been building up, or at least complicate it. Which of course is exciting, but exhausting. Especially when you've already had to go back over things any number of times to accommodate new observations. New to you, that is, but perhaps made hundreds of years ago. I have long been familiar with Kantian idealism of course, the notion that our senses can perceive things structured in the way that suits them and us, hence intercept and shape a world that is peculiarly *our* world, not the real, or as Kant calls it noumenal world, about which nothing can be said. It's a depressing formula in that it simultaneously declares that an absolute reality does exist, but then shuts us off from it in an 'ideal' space of our own. What it has going for it is its understanding, which most people seem to repress, that absolutely nothing can be known to us that does not, in one way or another, come through our own senses. But how could it be otherwise?

The Spread Mind starts from the same position, but declares that the world your senses carve out simply is your experience, and *is* reality, though not of course the whole of reality.

I was aware too of the rather reductive linguistic version of Kant's position that you come across in the twentieth century: since we can't think without language, certain philosophers claimed, language structures the way we think, and the way we respond to perceptual experience. Hence language structures the only world we can know, while again, what is absolute, or simply beyond our language, remains out there sovereign and unknowable. This too is depressing, but entirely in line with the human vocation for collectively isolating ourselves, constructing a world out of words and thoughts. Still, this world too has its performed reality, words and world meshing in experience, and in any event, a little quiet observation, or some training in meditation, and it is quickly evident that we can have all kinds of perceptual experience that lie outside any structuring in language. Our talking to ourselves or to others is not everything we experience, and though my name for an object, or experience, where I have one, may shift or 'colour' my perception, it does not constitute my perception *tout court*. I see an apple pretty much the same way whether I'm speaking English or Italian.

What I was not aware of – to arrive at my point – was Husserl. So that discovering him – Edmund Husserl, 1859–1938 – in the summer of 2016, I had to weigh very carefully the conclusions he reached about perception, consciousness and objects, with those of Manzotti, because for quite a while the two seem to travel the same road, then part company. Who is right?

You may complain that anyone writing a book about these matters, any matters, should first read everything there is to be read, digest it, and only then set down his own opinion. This would be all very well if one could ever imagine reading everything that has been written, or indeed is being written as we read; and if we could ever imagine retaining the huge quantities of information, opinions, theories and reasoning we would come across. And then retaining them in any stable way, since everything we read will alter our response to what we read before, and in any event our memory of what we read tends to change with time, so that if one goes back and rereads a text, it may well seem rather different than the idea we had formed of it, or rather have been forming of it ever since the moment, or period, of reading. Nothing stays still. So while we may feel we know more and more about our subject, we never feel we have *arrived*, or that we would not know even more if we simply *reread* everything we have already read, including perhaps the things we have written ourselves, which all too soon begin to fade into the past along with everything else. It's a struggle.

Hence if things are going to be written, the only way to proceed is to decide when one is ready and write. And for me when I am ready is when there is a balance between having read enough to feel reasonably sure I am 'in the zone', as it were, and having enthusiasm enough, naivety enough, to embark on a large project. And having enthusiasm means feeling that you can write something in some way *different*, whether in content, or approach, than other things that have been written, something *worth reading*. In short, when, rightly or wrongly, you feel you have something to say. Even if it is only something about your involvement with the problem, rather than a final solution.

Does this feel like a pointless, self-regarding digression? It is not; it is right in line with the sudden interest in Husserl that flared up a couple of months back. For precisely the insight of Husserl's that most intrigued me was his sense that not only, as in Kant, is conscious perception structured by our senses, but that part of that structuring is our constant impetus to acquire more and more experience of an object, building up a more and more complex picture of it, our expectation that an object seen from this side will also be seen, though perhaps rather differently, from the other side, that if cut in half it will have an inside of this or that kind, that if seen under a different light it will change colour in this or that way, that if left in the sunshine it will grow warm or perhaps rot. Etc.

Like the enactivists, Husserl feels we are in a constant process of *intentional* engagement with the world of objects, what Fuchs called *Erleben*, or conation; it's the kind of process you see described in Woolf's *Mrs Dalloway* as Clarissa stands on the pavement and takes in the street scene in Chelsea: 'For Heaven only knows why one loves it so, how one sees it so, making it up, building it round one, tumbling it, creating it every moment afresh.' Obviously this 'making it up' aspect towards experience could equally well be applied to someone trying to understand something like consciousness.

However, and here Husserl is very much in line with Kant, he doesn't think we can know anything about the *absolute* object, which he nevertheless is convinced exists. There is a real thing out there quite separate from our experience of it, but when it comes to knowledge all we can do with that real thing is 'bracket it off' and concentrate on the phenomenal world of our experience; it was this 'bracketing off' approach that led to the so-called school of phenomenology, of which Karl Jaspers would later be part, and later still Thomas Fuchs. Consciousness was to be studied as a real thing, but as it were separate from the things of which it was conscious, separate to the point that, rather surprisingly, Husserl believes that consciousness can exist entirely *removed* from the world of objects, without the physical at all, and this of course, is in direct contradiction with the Spread Mind theory, in which any consciousness implies an object with which it is one. The consciousness is its object; so, no objects, no consciousness. But on reflection it is impossible to imagine a world without objects, since our very being conscious implies a creature who is him or herself a possible object of their own consciousness. There is always something.

I want to talk, then, in this last chapter, among other things, of this aspect of constant engagement with the world, this constant coming back to it (do we have any choice?), constant expectation of it, fulfilled or otherwise, in which experience happens, or which is one with experience happening, something that makes us think, maybe, of the functioning of those feedback neurons in Hannah Monyer's doomed mice, forever sending back impulses from the lateral entorhinal cortex to the olfactory bulb, causing the mouse's whiskered nose, one presumes, to focus more clearly, to investigate more carefully, or perhaps to drop one investigation for another; and actually it seems appropriate here that we all pause for a moment and think of a mouse when it's *not* anaesthetised and fastened to a laboratory bench with screws in its skull, but sniffing around something, the way mice do – a piece of cheese, an old potato, damp sawdust in the corner of its cage – pink nose raised in the air, or thrust this way and that, higher and lower, left and right. There is always something endearing about an animal sniffing around something, becoming fully acquainted with it, bringing it into being. And I want to suggest that this is not out of line with the notion of memory, or talking or thinking for that matter, as forms of *performance*. 'Every act of perception', says Oliver Sacks in *Musicophilia* – and here he is singing from the same hymn sheet as Virginia Woolf – 'is to some degree an act of creation, and every act of memory is to some degree an act of imagination.'

To create/make/perform are words with an obvious kinship. Crucially, they hint of a process that is incremental, and hence suggest that time as we experience it is not the instant, and not even simply (simply!) something spread or stretched, to bring together the whole of a sentence, this sentence, for example, this paragraph, if you will, or a musical phrase, or movement, or a goal scored on the break with three rapid passes and one sharp shot, but something *cumulative*. Creation means making something, or making it up, composing it, *through time*. Cumulative, then, but with always the possibility, and all too often the anxiety, of dispersion. Will we finish the performance before the thing begins to decay, before the desired object truly, that is fully, exists? Or rather, will it, the object, the event, properly *happen*, in a way that satisfies? Will the syntax hold? Will we get our mind round Mahler's Fourth? Will the goal be scored? It is not impossible to get lost in a sentence and not recall how it began, how it should go on. It is commonplace when writing a book, which is nothing if not a madly extended performance, to forget, or at least become uncertain, as to what exactly one has already said. Or meant to say. It's discouraging. And the same is true of reading, which is also a performance. The reader too can fail to hold together the syntax or the plot through to the end. Nevertheless you push on. Consciousness is forever in tension between the old and the new, what is being shaped and what is being forced upon it, a sudden idea, a power cut, a phone call. If your experience is the world, as Manzotti claims, the world is changeful, protean, kaleidoscopic; don't expect to be in control. Your experience could be an object, a building you're looking at maybe, across the street, and the object explodes. The object has radically changed. Your experience is changed, your consciousness utterly metamorphosed. Or, less dramatically, the sun goes in. The facade is different. The facade *is* your experience. Your body is here on this side of the street, your experience, or a part of it, there, on the other, where the light suddenly fades from the stucco, beyond your control, though carved out by your perceptive faculties, a dull, flat, dirty pink. But now the sun comes out again and the building is vivid again with all the sharp shadows of its balconies and the intense green of an ivy climbing the copper drainpipe towards jeans hanging to dry from a balcony rail where a man in a white vest leans out, smoking a cigarette, and salutes someone on the street.

Consciousness is all change, accumulation, dispersion, things that unexpectedly remain active, or repeat themselves, over years and years, a few words a teacher said at school, still very much in hearing range – 'All you're lacking is confidence, boy' – things you thought had gone but suddenly come back – the smell of a certain red sauce they poured on ice cream in your Lancashire infancy (sarsaparilla?) wafts by you fifty years later at a street corner in Milan – and things you imagined would remain, *must* remain, they hurt so much, or gave so much pleasure, and yet are quite gone, or so it seems; in fact there must be many such things you don't even know you've lost; you performed them once, then never again.

So though there is accumulation, in life, you don't control that either, or not as much as you would wish. You can't *hoard*. You can try; by performing, for example, the same words over and over, a poem, a formula you'll need in an exam, a password you really mustn't forget. You repeat it time after time, hoping it will cut a deeper and deeper groove, until the first word, or number, or verse, automatically produces the second, as one step the next in a waltz. A performance become a rote. But that too, if you don't keep performing it, will go one day. You take a three-week holiday and when you come back the password to the office accounts system is gone. Incredible. You can't remember it. You can't *perform* it. It's not in your head (but it was never in your head). Until the system gives you the password prompt, and there it is again, you are doing it, typing it, saying it, the way when you clip on your skis at the beginning of the new season, or climb on a bicycle for the first time in years, or pick up a guitar, you have the prompt, the performance can resume, perhaps a little shakily at the beginning, but it will all come back fast enough.

I say all this – and I confess I was a little carried away there, I hadn't planned to write so much – because I hope now, in describing two walks along the same path undertaken almost exactly a year apart, to fix more clearly in my head and yours, this notion that in fact there is nothing in our heads but neurons and that everything that seems internal and inside – our thoughts and memories – is in fact performed in contact with, in oneness (the body's contact, the mind's oneness) with, the objects of experience, near or far, as a dance is performed in oneness with a partner. And often the dance continues when the partner is gone, apparently. And sometimes, of course, the partner is still there, but you dance no more. Or it's no longer *that* dance.

So, after the neuroscientist's lab, the anaesthetised mice, the brain slices and the smell of formaldehyde, let's go back to the fresh air and a much happier laboratory, the wide open world, and back initially to September 2015. Following my encounters with Sabina Pauen, Thomas Fuchs and Hannah Monyer, having a couple more days to spend in Heidelberg, Eleonora and I set out to explore the town. In particular, we climbed the hill on the north side of the river, opposite the old centre, following a path known as Philosophenweg, Philosophers' Way. Not that there is anything particularly philosophical about the walk as such, but apparently the philosophers at the university in the early nineteenth century enjoyed walking and mulling things over here.

It was a cold damp morning. We had brought the hotel umbrella with us, though it wasn't actually raining. I had received an email overnight from my brother that had irritated me. I should say that another person I am in constant conversation with is my brother who lives in Red Hook, upstate New York. Across the Atlantic, we try out ideas on each other. He sends me photos of paintings he has made, being a painter by profession; I reciprocate with stories or essays. For months now we have been discussing Riccardo's theory, which he thinks interesting, but crazy, until this morning, or rather that September morning, 2015, Heidelberg – the 24th my old Outlook Calendar suggests – he had emailed me pages and pages of reflections setting out to refute Riccardo's notions and insisting that consciousness could only be explained in evolutionary terms as a manner of responding behaviourally to a challenging world. Above all, he had decided to support the position of Thomas Fuchs, whose articles I had sent him.

'Surely', my brother had written, 'the beauty of the enactivists' idea of a loop is that part of the loop is the object of experience and part of it is the biology doing the experiencing. The experience then is both the object and the biology functioning together.'

Why was I irritated by these earnest observations? Was it because this version of events was so close to Riccardo's earlier view of consciousness as process, as everything that goes on between object and brain, to the point that I felt challenged to return to that old idea? Or because I sometimes think that my brother takes an opposing position out of a mere spirit of contradiction, something my mother always accused my father of. Deliberate obtusity. We had already dressed for our walk when I asked Eleonora to hang on a moment while I fired off a hasty response. Fuchs, I wrote,

> suggests experience is a kind of *action* or, worse still, a kind of *knowledge about our actions*; this simply can't be true because – to keep it brief –
>
> 1. actions do not have *the properties of one's experience* (they are not red, cold, clammy, etc.), so . . .
> 2. actions are *no better than neural activity* when it comes to describing experience. They are not *the thing itself.*

As soon as I had sent this, I felt irritated with myself for replying hastily and ungraciously and also somewhat concerned that I had begun to write the way Riccardo writes, with numbered paragraphs. There was no need to win the argument at once. Or to win it at all, for that matter. There should be no question of win and lose in debates like this and one must always keep one's mind open to some new observation that shifts even the most tenaciously held ideas. But perhaps the real problem, I explained to Eleonora as we left the hotel – and I always feel more sane when I draw her into the discussion – the real problem was that I now felt uncertain about the relationship between this intuition I had of performance – memory as performance, thought as performance, talking as performance – which was actually very close to the enactivist idea of action, and Riccardo's conviction that the experience was one with the object.

We crossed the river on the Theodor Heuss bridge, barely a stone's throw from the hotel, and there was a light mist on the water and a sculler twisting his head to make sure his next stroke would take him under the arches beneath us. Was his experience, I asked her, the action, or the sight of the approaching bridge, the cool air on his bare arms, the twisting muscles as he turned his head?

'I can't see the problem,' Eleonora remarked. 'The action is required for the experience and is itself reaction to experience, but it's not the experience itself. You need to create to have a creation. But the creating is not the creation.'

'So the experience is the boat, the water, the bridge – or those bits of each he's in contact with – the oars, of course, his body as he sees and touches and hears and smells them all at once.'

'The damp air, the clouds. Plus something else he might be thinking about,' Eleonora observed.

'A hot shower.'

'Could be.'

'Something he has to pick up from the supermarket for lunch.'

'Why not?'

'But not the neural fizz.'

'We've said a thousand times we don't experience the neural fizz.'

271

There are time when Eleonora, who has only been exposed to the Spread Mind in conversation, seems to have a better grip on it than I do who have read a stack of articles.

'When did anyone experience their neurons?' she laughs. 'Right.'

This seemed fair enough and calmed me down a little. But the fact is, things are never settled when you take on these subjects, or even when you look at a boat sliding under a bridge. Husserl was right about this. It's as if you were always shoring up a sandcastle against the tide. This view of things. Nothing ever feels quite solved.

We studied the map they had given us at the hotel reception desk. We must proceed a couple of hundred yards beyond the bridge then turn right. Philosophenweg was narrow and surprisingly steep, and we were frequently having to stand aside for passing cars, the drivers smiling in entitled acknowledgement, wealthy folks by the looks, with pleasant houses perched on the hill, among rich vegetation and picturesque stone walls. After five or ten minutes, the effort of climbing so sharply warmed us up and I have the very distinct memory, right now, as I write, of the moment I decided to loosen my scarf (though I can't remember which scarf I was wearing), to let some cool air into my clothes. Then at a certain point the slope levelled off a little and to the right the houses gave way to a garden that dropped down towards the river now surprisingly far beneath us with the steeples and domes of the town beyond, hazy through the mist.

At the entrance to the garden – because we thought we'd go in and sit a while – was a tall stone remembering a certain

Joseph von Eichendorff, who had lived from 1788–1857; sixty-nine years of consciousness gathering and dispersing. We stopped to consider a life-size brass bas-relief in profile of a young man's head with a pointed nose and hair brushed forward horizontally over his ears in the style of the time. We would have assumed he was a philosopher, this being Philosophenweg, except for the fact that below the bas-relief there was a poem engraved in the stone:

Schläft ein Lied in allen Dingen,
Die da träumen fort und fort,
Und die Welt hebt an zu singen,
Triffst du nur das Zauberwort.

Of course, I didn't type this out from memory. Nor did I take a photograph. I found the lines on the Net. But I remember we read them together that day, after the steep climb, and a year later I would read them again on the same stone, or create them there again, when I went up the same hill on my own.

'Tell me what it means,' Eleonora challenged my German.

'I'm really not sure.'

'Try.'

I drew a deep breath. My technique, if we can call it that, when translating something I don't initially understand is just to launch in and see what happens.

'*Schläft* means "sleeps". *Lied* is "song" as in Schubert's *Lieder*, which is the plural. *Allen Dingen* is "all things".'

'Sleeps a song in all things,' Eleonora said.

'Genius.'

'Go on. *Die da träumen fort und fort*.' She pronounced it as if written in Italian.

'*Die*, pronounced "dee" is usually the definite article in the plural, but here, since there's no noun attached, it must be the relative pronoun attaching to *Dingen*. "All things, Which". *Träumen* means "to dream". *Fort* is "away", in the sense of not here. You remember Freud's essay on the "*fort da*" game where he observes a tiny child playing in his cot. Pushing a toy away, the child says *fort*, then recovering it he says *da*. Exploring the world and his control of it. Fuchs's territory.'

'And "*da*", "*Die da träumen*"?'

'I've no idea. From?'

This stumped us. How were the words to be put together?

'If we had a smartphone we could check it, or just find a translation on the Net.'

Eleonora is constantly pointing out the advantages of smartphones, but won't buy one herself until I have come on board. I think they will just lead to us being more distracted. A year later I would realise we had both been right.

'Surely more fun to translate it ourselves, then compare it later with something professional.'

'OK,' she says. 'So: "Sleeps a song in all things, / Which something dream away, away".'

'What about, "on and on"?'

'Sleeps a song in all things, / Which do dream on and on.'

'We're inventing.'

'It's a performance,' she observes. 'Next words please.'

'*Und die Welt hebt an zu singen, / Triffst du nur das Zauberwort.*'

'Funny,' Eleonora observes, after I read, 'how we can hear the rhythm and rhymes without knowing what the words mean.'

'Well, we know that *Welt* means "world", as in *Weltanschauung*. But I don't know *hebt*. Must be a verb. Third-person singular. *An zu* usually gives the idea of starting to do something. Beginning, maybe. *Singen* "to sing", obviously.'

'And the world starts to sing.'

'Could be. Though you would have expected the verb *beginnt*.'

' "Sleeps a song in all things, / Which do dream on and on, / And the world begins to sing". *Triffst du nur das Zauberwort*. Reminds me of the *Zauberflöte*.'

'Or the *Zauberberg*.'

'Magic flute, magic mountain, magic . . . word?'

'Easy.'

But the first part of the line is a struggle. I can't understand *Triffst du nur*.

'Take it slowly,' she says.

'*Nur* is "only" and *triffst* is the second person of a verb. I think it's *treffen*, but I don't know what it means. So *Triffst du nur* should be something like, "you *treffen* only", or, since the Germans are always inverting things, "only you *treffen*".'

'If only you *treffen*?'

'Brilliant! "And the world begins to sing, or will sing, if only you *treffen* the *Zauberwort*".'

'Almost there!'

Eleonora recites: 'Sleeps a song in all things, / Which do dream on and on, / And the world starts to sing, / If only you ... speak/guess/invent/find ... the magic word.'

'Lovely!'

'Romantic.'

'Isn't it bizarre we had less trouble with a poem written a couple of hundred years ago than with a stupid menu?'

'Because the words are all basic things you learned at school.'

'Right. But I could never have come out with them, or put them together in any sensible way, without seeing them written. We jumped from one to the next like stepping stones and they perform the poem.'

'You're obsessed with this idea.'

'At least I don't believe there are magic words that will start the hills singing and the trees clapping their hands.'

What a charming idea it was, though, we decided, proceeding into the garden, that there was a song in all this countryside, the woods rising up the slope behind us and the river flowing under the mist below, a song in every stone and twig that would burst forth, if you only you could come up with right abracadabra.

'Frustrating, though, since one never does come up with it,' Eleonora observed. 'Romantics always yearn, without the great revelation quite happening, except maybe in some flash of epiphany that's gone before it's grasped.'

It struck me that this attitude was the opposite of philosophising, of working away at the facts, coming at them again and again from different angles with your nose alert to any scent or trail. Romantics wanted meaning to suddenly shout itself from the treetops.

'Ye Presences of Nature in the sky / And on the earth! Ye Visions of the hills! / And Souls of lonely places!'

'I beg your pardon!'

'Wordsworth. *The Prelude*. We learned it by heart at school.'

'Sounds like panpsychism,' Eleonora thought. 'What do we think of panpsychism?'

'Not much. It's as if, unable to locate consciousness or grasp what it is, you simply decide the whole world is conscious. And if you could find the magic word it would all speak back to you.'

'You're being reductive.'

'I can understand the people who believe that the world *as a whole* is, as it were, alive, in the sense that it forms an all-encompassing, interconnected, living system, but how can you believe that a stone is conscious?'

'Isn't that exactly what Riccardo is saying when he says the object is the experience? So if you're looking at a stone, the stone is your experience, so the stone is conscious. It begins to sing when you look at it.'

'No, sorry. The panpsychics believe the stone is conscious *without* my looking at it. Riccardo's hypothesis is that when I look at a stone there is causality between body and object such that an experience occurs which he locates at the stone, but happening there only because of the presence of my body. The object/stone is my experience, but not some Galilean absolute stone, a Kantian noumenon; my experience is not the crystals or defects inside the stone, or the damp surface or the invisible part on the other side. Just what I see, or maybe touch, whatever.'

'OK, OK,' Eleonora laughs.

We've sat down on a bench in the garden, which is laid out in rockeries and flowerbeds between banks of autumnal shrubs. It's all wonderfully melancholy and quietly damp, not singing at all, but infinitely more congenial of course than a lab full of drugged mice. The only other visitor is a man with a large white dog that is peeing against another memorial stone, dedicated, as we will discover in a few minutes' time, to Friedrich Hölderlin, a pantheist who believed nature was not only conscious but divine and that ultimate meaning could only be grasped, or announced, in poetry; certainly not seen with a laser microscope through a window in a mouse's skull.

Eleonora asks if there are any serious panpsychists around, and I tell her, many. In fact, I had recently read an article by Galen Strawson upholding exactly this position. I had been rather surprised, I remembered, to find Galen Strawson, whose accounts and criticisms of other people's work are wonderfully clear and elegant, supporting this position, which he called *physicalist* panpsychism. But he does explain it very elegantly.

'The argument is simple. Everything that concretely exists is physical. All physical phenomena are forms of energy. Hence everything that concretely exists is energy. The intrinsic nature of energy is experience, consciousness. So everything that concretely exists is conscious, or a form of consciousness.'

'Why should energy be consciousness? Who says?'

'Indeed. I think the argument goes that everything must be in a relation of physical cause and effect with something else. Everything is *caused* by *something* that came before it according to the laws of physics. That is: there is nothing that *emerges* magically, that isn't *caused*. Hence consciousness must be a feature of concrete reality. Hence consciousness is a feature of everything.'

'But why couldn't it be a particular manifestation of energy, like a stone is one kind, a tree another?'

'That's exactly what I thought. Strawson is a panpsychist without being in any way a mystic. So basically what he's saying is: not enough is known about matter for us to say that it doesn't have properties for the moment obscure to us. So everything is up for grabs.'

'You can have your magic without being magical.'

The sun was coming through the mist now and the autumnal colours of the trees all around us were beginning to glow. I sometimes wondered, I said, whether the whole point of panpsychism wasn't to surrender yourself to a spirit of magic and instant total meaning. What the world will start singing when you find the magic word, is the meaning of life, the universe and everything. To call it physicalist was just a way of attempting to give seriousness to the move, of saying magic is real.

But Eleonora disagreed. Eleonora had done some research into Romanticism at the university and her impression had been that Romantic yearning was not a desire for instant meaning, but for kinship, as in the 'Ode to Joy', *Alle Menschen werden Brüder*, or in the Pastoral Symphony where nature becomes music. The desire was to feel part of the world, part of nature. So making everything conscious, you made everything potentially friendly, or at least company, ready to burst into song. Ready to speak to you.

'It's a reaction,' she explained 'to the rationalist separation of subject and object. You feel frustrated, even angry that you're not really *in* the world. You're alienated. Things like pantheism and panpsychism are a way of getting back in. Astrology is a nuttier version. You're connected to the stars. Involved in the universe. Even those people who say we all live in a virtual world on information networks are really trying to see how we can feel connected with everything.'

We stood and went to look at the memorial stone where we read Hölderlin's name. He had been connected with everything from 1770 to 1843. For some reason I had imagined he had died young. There was another poem, but this time we didn't get past the second line, it was too difficult, and so, frustrated by this failure to have the experience happen, but laughing that we were able to describe it that way, we left the garden and followed the road which ran across the hillside, drifting slightly upwards into the woods. After five minutes the asphalt gave way to dirt track and after ten there was a fork, one path climbing steeply up to the left, the other dropping back down into the valley towards the river. We pulled our map out and discovered that we were no longer on it, so to speak. It stopped just beyond the garden. Another reason for having smartphones, Eleonora observed; you were always on the map with a smartphone. We pushed on for a while, climbing higher and higher through thick woodland, hoping for some signpost, to a village perhaps, where there might be a place to have lunch.

As we walked, I began to wonder, without saying anything to Eleonora, when would be the appropriate moment to turn back, especially because it was getting cold and had come on to drizzle. We opened the umbrella. But we only had one. I hate turning back. But equally I hate walking when I'm wet. The trees were majestic on either side of the path, wonderfully woody and mossy and earthy. But somehow unpromising, recalcitrant, not easily part of some theory or thought pattern, or discussion of Romantic poetry. The air in particular was damp and dense and laden with nature. Air is quite different when it moves among thick foliage over wet mud. Consciousness is different with it. We were moving through a vast invitation to quietude, stillness. The place didn't so much brood as do its living noiselessly, without fuss, without yearning. It was a place you felt you might once have come from, but couldn't easily go back to. Or not if you wanted to still be yourself.

Then Eleonora asked me what I was thinking and I told her I was wondering when to turn back and she said she had been thinking the same thing, so that very briefly we considered the difference of words thought silently and words pronounced out loud in conversation. Were they in the end the same phenomenon – performances leaning on strings of words – the only difference being the physical vocalisation? We still hadn't quite resolved this when suddenly the rain turned heavier and that settled it. We turned back. Almost at once, hurrying down the slope now, we met a group of girls, kitted out for a serious hike, climbing up through the trees towards us in colourful waterproofs. I asked them did they speak English, and they said they did. Was there a place for lunch nearby? They pointed up the hill and mentioned the name of a village, a name we immediately forgot, because of its Germanness, because it found nothing to cling on to among what we already knew. How far, we asked? Shoulders were shrugged. There was some giggling. An hour?

We opted for the return. A few minutes later, branching off from the Philosophenweg where a path promised to plunge steeply down towards the river, I was aware of the possible analogy between our morning's walk and the way perception works, perhaps the whole business of philosophising too, exploring, pushing on, trying to establish a goal, dreaming of the magic word that would make sense of everything – *lunch*, perhaps? – finally turning back to the comfort zone of the already known, or at least familiar, hopefully dry. But an analogy, in the end, is just another prop that presents itself, another path opening up in this and that direction. And no analogy is ever exactly the direction you planned to go in. It's never perfect, otherwise it would be identical with its referent, the thing you're struggling to describe. Every analogy is a deviation. But then what can you do but follow the paths that present themselves along the Philosopher's Way? Eventually, we got down to the river quite a long way out of town and had to walk half an hour in heavy rain to get back to the hotel.

So that was 24 September 2015, and I would return to Heidelberg and do that walk again on Saturday 29 October

2016, very shortly before writing this chapter. The Deutsch-Amerikanisches Institut had tried to get the participants in their Science and Religion project together for the spring of 2016. And then for the early summer of the same year. But it was impossible. Everyone's life was too connected elsewhere to find a moment when we could all come together. Until 28 October. Then there was to be a single evening in which nine of the ten authors would say their piece about science and religion.

And the tenth?

Raoul Schrott had the whole evening of 27 October to himself in what the institute's website, which I ran through Google Translate, described as a 'pre-taste' that 'strikes an arc from cosmos to microbiology'. Comically, the German word *Schrott* means scrap, or scrap metal, and since of course even Google Translate's new and remarkable translating program is still only a program, not something that has ever lived or knows what it's doing, or what words *mean*, and since, in German, nouns like names take capital letters, the service has no way of distinguishing between the intellectual, Raoul Schrott, and the waste product. So we hear that 'In varying poetic forms, scrap deals with today's knowledge about the world: from the Big Bang to the origin of the earth to man.' The computer doesn't know that only people not scrap could do this kind of thing. I should say that since my own surname spelt backwards is skrap, I felt a special interest in this mistake. In fact, when writing comments on the Net, I often sign myself 'skrap'.

Did Schrott have an evening to himself on the 27th because he had worked particularly hard on the project and had something particularly interesting to say about science and religion, or simply because he had a prior engagement for the 28th? I have no idea. We were not told. The book the Deutsch-Amerikanisches Institut had put together from the pieces we had all written for them came only in German, so I couldn't read Schrott's piece. The title was *Wissenschaft – die neue Religion? Literarische Erkundungen* (Science, the New Religion? Literary investigations). I later asked the institute to send me the originals of the pieces written in English, by Colm Tóibín and the American writer Ben Marcus and of a piece in Italian by the Italian novelist Michela Murgia. The three texts, when they arrived, could not have been more different from each other, yet one thing they all demonstrated, something amply confirmed too on the evening of the 28th, was that dualism is alive and well; most of us are still happy to divide the world into the realm that belongs to science and the realm that belongs to the soul, or to literature. Far from being a religion, for these writers science presented itself as a threat, while poetry, prose and prayer, or simply any protestation of the sacred supremacy of the individual's secret life, were all forms of noble resistance.

First there was an aperitif, to be drunk on our feet, milling among tables with tiny munchies while smartly dressed waitresses offered sparkling wine. I took the orange juice, since I realise that my perceptive faculties carve the world out rather differently when in company with alcohol; words in particular are intersected and arranged rather differently and with potentially embarrassing consequences. So I've learned never to drink before any kind of public speaking, however apparently 'easy'. Colm Tóibín, whom I was meeting for the first time and found as rumbustiously merry as his writing is elegantly melancholy, had no such qualms and, drink in hand, was expounding on the literary genius of Bob Dylan, announced two weeks before as winner of the Nobel Prize in Literature. *'For one thing that's certain'* – he hammered the song from *Nashville Skyline* – *'You will surely be a-hurtin', If you throw it all away, ay ay ay.'* Tóibín flourished his drink. 'Lo-o-ove! Isn't it wonderful? Great rhymes. *A-hurtin!* So simple. What a poet!' He sent up the song, *'She said she would always stay ay ay ay.* You see how subtle that line is? *But I was crool, I treated her like a fool, I threw it all away ay ay ay. I threw it all away.'* He laughed drinking his bubbly, and I thought that funnily enough this was the theme of so many of Tóibín's stories: lost love.

'Wasn't there perhaps,' I wondered, 'some irony in the way Dylan sang that song?'

'Not at all.' Tóibín shook his head. 'Not at all!' At which point I had to wonder if *he* wasn't being ironic. There is no scientific instrument for establishing the presence of irony; it's something we decide among ourselves. Ben Marcus joined us, the conversation stayed on love, and Tóibín was now remarking that of course people always did throw things away, however many songs you wrote to warn them, because curiosity for the new was always stronger than love for the old; when some bright young thing came along you could never say no. Could you? I smiled and told him I would have agreed with that ten years ago, but that right now no new thing however bright and young would budge me from my lady in Milan. I simply wouldn't dream of it. 'Lucky you,' he said, eyeing me suddenly seriously. 'How sweet,' he said. For the first time I glimpsed a connection between the man and his books.

Now it was time for the main event which took place on the floor below, a large low-ceilinged room with a low stage up front and seating for maybe 150 people. The authors – four Germans, an Italian, an Irishman, an Englishman, an American and a Russian, that is to say six men and three women – were seated together in the front row. Jakob Köllhofer, he who had offered me a considerable sum of money, gave a general introduction, explaining how and why the Science and Religion project had been organised, how interesting and edifying it had been, and so on. After which, Professor Jan Wörner, again male and German, head of the European Space Agency, gave a 'keynote speech', standing at a lectern with a fancy PowerPoint projected onto a big screen behind. But why don't I just give you the same programme I held in my hand and was now studying with sinking heart:

	Programme	Who	Duration
19.50	Percussion	Klaus van Boekel, Vera Matenaar	10 min
20.00	Welcome and introduction	Jakob Köllhofer	10 min
20.10	Keynote: Points of contact between religion and science	Prof. Jan Wörner, European Space Agency (ESA)	20 min

	Programme	Who	Duration
20.30 bis 21.15	1st conversation *Die Magie des Anfangs – German (The Magic of Beginning)* *(following: percussions)*	**Daniel Kehlmann, Michael Maar, Kathrin Passig** Mod.: Thorsten Moos, FEST	45 min
21.20 bis 22.10	2nd conversation *Intimacy and Intuition – English* *(following: percussions)*	**Michela Murgia, Ben Marcus, Tim Parks** Mod. Russ Hodge (science writer)	45 min
22.10	break / percussions		10 min
22.20	**Theatre play (extract)** 'Und wann kommen die Elefanten?'	**Judith Kuckart**	15 min (DVD)
22.35	3rd conversation *Heading for a new metaphysics? – English* *(following: percussions)*	**Viktor Jerofejew, Colm Tóibín** Mod. Russ Hodge	45 min
23.15	Book presentation and closing remarks	Jakob Köllhofer	10 min

Why was my heart sinking? It began during the drumming and keyboards which had preceded the event proper as we all settled into out seats. A white man with dreadlocks, presumably the Klaus van Boekel announced in the programme, dressed in vaguely African fashion – perhaps he was South African – played bongo drums beside a lithe girl on keyboards. The music was insistent and although I am a sucker for almost all manifestations of live music, I had the immediate feeling that this wasn't something I would have chosen to listen to. And in general I would never have *chosen* to be here, trapped on my seat in full view in the front row. But, looking at the programme, I saw that nevertheless, like it or not, I was going be here, in the best-case scenario, for at least the next three and a half hours, and very likely much longer, since all my experience of public speakers tells me that they never never never keep to time.

The keynote performance was not encouraging. Jan Wörner, no doubt a brilliant man in his field, had nothing to say about religion. Science and religion were absolutely separate, he said. They moved in different realms according to different criteria. Likewise the people who interested themselves in such matters, he seemed to suggest. There was an irritating complacency, even self-congratulation about his delivery, reinforced by a slick PowerPoint full of dramatic video clips on the history of the universe; apparently everything was more or less known, everything more or less under control. Indeed it was hard to understand why, in this case, anyone would have wanted to organise such a conference. Or, alternatively, if our keynote speaker was wrong, why anyone organising such a conference would have invited someone so obviously out of sympathy with its inspiration to deliver the keynote speech. What was required was an anthropologist, not an engineer.

Wörner was speaking in German, so that I was listening to an interpreter's voice in a headset, constantly aware of the man's boyish haircut on a middle-aged face, and his energetic, strongly emphasised delivery, in German, which could not have contrasted more sharply with the rather cautious woman's voice giving me his words in a deadpan English. It was as if I were being invited not to take what he said at face value. Translation can do this. In particular interpretation. Speakers seem less convincing, simply because the voice is split from the speaker.

In conclusion, Wörner wished us all a rewarding evening, apologising that he wouldn't be able to remain with us as he had a prior engagement. Science had spoken, it seemed, and was not interested in what literature had to say in reply. Separate worlds. Consulting the programme again, I now realised that the session I was speaking in was entitled 'Intimacy and Intuition', though neither the article I had sent to the institute for inclusion in their book, nor anything I planned to say, seemed really to fit this theme. Nothing had been discussed or prepared. Would the first session, I wondered, eager for guidance, stick by its title – *Die Magie des Anfangs* (The Magic of Beginning)?

No, it wouldn't. After the ten minutes of percussion, the ten-minute introduction, the twenty-minute keynote speech, we now had forty minutes of discussion, where four voices, the chairman and three writers, were again all alienated by the deadpan delivery of the hard-working interpreter. Looking back now, I can remember almost nothing about this debate. Perhaps I was in the wrong mood. For a moment my mind was distracted by the thought that although *Zauber* in

Zauberberg obviously meant magic and was translated into English as magic, yet the Germans had the word *Magie* which also translated as magic.

What was the difference?

All three participants, a woman and two men, told the chairman that their stay in Heidelberg, their meetings with scientists of various disciplines, astronomers, quantum physicists, neuroscientists, had been extremely intense, indeed had *changed their lives.* This perplexed me. Did they mean it? Were they aiming to please their hosts? Was it a translation problem? At some point there seemed to be a bit of a tiff between the woman and one of the men over an issue that had to do with gender. But the fact is that as I was struggling to follow this debate, another thought began to assert itself, an idea that my mind immediately attached to: I must use this moment I was experiencing, this sense of entrapment, of wishing I was somewhere else – something I suffer from all too often at conferences and public events – use it to talk, in my book, about the question of *intention*, its role in the consciousness debate, and its relation to dualism.

Husserl followed his mentor Franz Brentano in making intention or intentionality a defining characteristic of conscious experience. In our day Daniel Dennett would pick up on this. We're not talking here of intention in the everyday sense of planning to do something: the minister intends to raise interest rates; Margaret's one intention was to get drunk, etc. Or not only. That kind of intention does come into the picture but, at a more basic, moment-by-moment level, we're suggesting that consciousness is the product of the mind's reaching out to the world, intending, grasping it, being *about* it. Some philosophers would even start using the word 'aboutness' to describe mental activity. It is always *about* things, without being the things it is about.

Looked at long term, the idea of intentionality goes back to Anselm of Canterbury in the eleventh century who distinguished between objects in the understanding and objects in the real world; the first were the product of the mind's intention, or intentness towards the second. In *Being and Nothingness* Jean-Paul Sartre would make intentionality synonymous with consciousness; John Searle would say consciousness was 'intrinsically intentional'. This belief is in tune with a vast literature from many cultures in which mental life is presented as a kind of appetite that *consumes* the outside world. Some of India's pre-Vedic thinkers, way way back, thought of the world as conjured into existence by the mind's appetite. In Leonardo da Vinci's anatomical sketches, the eye is sometimes shown as a mouth devouring the world. More recently, of course, this metaphor has got mixed up with the IT analogy so that we now speak of the brain as devouring 'information', 'number crunching', and so on.

Once Husserl had established his central concept of intentionality – such that the real world is separate and un-knowable, bracketed off beyond discussion, while mental life is simply intention towards that reality – then his writing breaks up into a never-ending list of categories and fine distinctions as he tries to put in order all the different ways a thought or perception can be intentionally related to an object – appreciation, judgement, desire, rejection, etc. – depending to an extent on the kind of object it is, whether it is actually present, or if it perhaps never existed at all – angels, demons. This is taxing reading and remarkably similar on the one hand to many Buddhist philosophical texts, equally concerned to nail down the exact nature of different kinds of thought experiences (in particular whether they are pure unmediated reflections of the world, or contaminated reflections, that is perceptions coloured by the mind's intentionality), or to Hume's classifications of every kind of thought and perception, or even to Gilbert Ryle's *The Concept of Mind*, which again tries to say exactly what an endless series of 'mental states' amount to.

Whatever the metaphysics behind these discussions, you marvel at the intricacy with which experience is described, but at the end of the day you never feel the argument can *go* anywhere. It can't say anything about what consciousness actually *is*. Certainly, in the case of Husserl and other intentionalists, you soon realise that the whole approach is predicated on the separation of subject and object and the consequent paradox that things seem to be simultaneously outside me and inside my head. How can the room I am sitting in, William James put it, be both out there and in my brain?

The idea of intention, aboutness, then, is an attempt to bridge that subject/object divide. The subject is separate from the object, but *about* it, intent on it. Your consciousness is *aboutness* of the object.

But does aboutness really exist? Or intention in this sense? Does it mean anything, or tell us anything? Would the world disappear if I stopped 'intending' it? Am I able to stop? And if I'm not is the word 'intention' drained of all sense? Is such a disappearance what the Buddhists mean by *sunyata*, the state of emptiness one reaches (sometimes) in meditation, where it seems one is contemplating a strangely reassuring nothingness? Otherwise known as the not-self?

Sitting through the debate at the Deutsch-Amerikanisches Institut, I began to wonder how this notion of intentionality could be applied *to this very moment*. Can you be said to be 'intending' the things you perceive if you would much rather *not* be perceiving them? I was certainly intending other objects of thought, i.e. thinking about other stuff, most of all this reflection on intention, a reflection I was trying to develop (making it up, building it up), as it were, in resistance to the present situation, in order to have the time pass without suffering this debate, which seemed irretrievably dull. Nevertheless, however valiant my attempts, the debate was imposing itself upon me, I wasn't altogether succeeding in resisting it, and of course that was quite simply because I was *there*, sitting in the front row with a headset on my head, not to mention Colm Tóibín equally restless, I suspected, to my right.

Husserl's approach, as I hinted before, has this in common with Riccardo's, that since perception is constantly altering focus, experiencing an object from different angles under different lights in different ways, time becomes a rather complex issue, a constant stretching and gathering, since my knowledge of the object, like my engagement with a sentence or a piece of music, builds up, changing and fulfilling or not fulfilling its initial promise. Each instant is built on the last, integrated with the last, holding together the immediate perception of the object, perhaps a cathedral facade in sunlight, but with the knowledge of the facade in shadow, or the cathedral from inside, or seen from an aerial photograph.

But the key difference of course, is that while Husserl claims that experience is not the real world but a mental representation, Riccardo denies the subject/object divide and insists that your experience simply *is* the world. I, for example, at this moment, am not 'intending' the dull debate in the Deutsch-Amerikanische Institut – the stage, the participants, the big screen behind them (now projecting the question *Wissenschaft – die neue Religion?*), the disorientating confusion of German voices out there and this monotone English in my headset, my slight discomfort in the chair, a vague need to go for a pee, a certain anxiety as to what I will say myself when the time comes, and so on – I am not intending these things, I am not *about* them. My experience *is* them. It is located where they are located. My body is here, on the chair, my experience, or certain elements of it, there, up on stage. And there and there and there. The participants, the moderator, the screen, the backdrop, the lighting, all come into being, for me, thanks to their encounter with my perceptive faculties. The same being true for everyone else in the room.

In this version of events my unease with the situation will have to do with the resistance of certain elements (of me!) to accept identity with others. In a train when someone sits down beside me and begins a long loud conversation on their phone about an unfaithful husband or a violent child, I quickly get up and move into the next carriage. I don't want to *be* this conversation. I don't want identity with this anger or angst. When I start to watch a film that gets seriously ugly, or seriously dull, I turn it off. Or leave the cinema. The accumulation of past experience, the inertia we call me, reacts against this imposed identity.

Sitting through the final interminable minutes of that debate, Riccardo's theory made perfect sense to me. There was a struggle to focus on distant things – this question of intentionality, certain authors I had read, Husserl and Sartre, Searle and Dennett, my correspondence with my brother, this book I am writing – an attempt, you could call it, to renew, or, against the odds, to privilege the causal relationship of these past 'events' with me now, in order not to be present in the here and now, not to suffer identity with the debate. But I couldn't do this entirely. Because of course the German on stage and the voice in my headphones were the here and now. My body was where it was. And because in a few minutes I would be up on stage myself.

So you'd better concentrate, Tim.

And concentrate I did. The time came for the previous speakers to step down and for our group to step up. I performed my part. I said my bit about the ways in which science occupies some of the same spaces religion used to: its claim to offer an account of how the world came to be, of our place in the universe, its pretence of superior authority, its vague promise of a better future, perhaps even some day of the chance of downloading our minds onto some supercomputer and becoming immortal. Before me, Michela Murgia spoke as a devout Catholic about the religiosity or otherwise of the scientists she had spoken to and the ethical problems of the research they were doing. After me, Ben Marcus spoke of the marvel and menace of the immune system. It was all rather inconclusive and very likely just as dull as the forty minutes that had preceded it. Certainly, nobody had anything to say about intimacy or intuition. At the end of my own spiel I suggested that while religion offers a formula now that holds out hope for the future (life after death), science offers hope now of having formulas to offer in the future (all kinds of ways of improving and extending our lives, if not quite immortalising us). This, I hoped clever formulation of my own, roused no interest or assent whatsoever.

But the second half of the evening improved and was at last amusing when Tóibín took the stage and told a hilarious anecdote of how his resistance to science began with his chemistry teacher's attempt to court his widowed mother, something he had experienced as a terrible threat to his domestic happiness. He went on to speak of black holes, which quickly became a metaphor for a certain state of mind, or social miasma, and concluded with some reflections on the relation of German scientists to anti-Semitism. He was eloquent and engaging and nothing he said seemed to have required a meeting with scientists in Heidelberg to have said it. Later, reading the piece he had submitted for the project's publication, I was struck by the implications of its climactic moment. It takes the form of a short story rather than an essay, but involves a novelist who is invited to Heidelberg by friends of his now dead scientist father to talk to leading astronomers and physicists. Their account of the vastness of the universe leaves him concerned about 'his own uselessness'. In response he goes back to his room and 'began writing as though he had some desperate need to reassert himself, take back control of narrative structure from those who would attempt to suggest that the small detail, the very idea of a single thought or a memory, were nothing against the sheer vastness of time and space and theory'.

Attempting, then, to reassert the significance of the individual, he writes about a man alone looking at the sea (something Tóibín has already done very beautifully in the story 'The Empty Family'). The man reflects, in melancholy fashion, on the act of looking, the world's mutability, the mutability of memory, the sense of consciousness as flux. Eventually, the novelist feels that this writing 'had somehow freed him from the oppression of the day, from the world of calculation, from a world in which knowledge had a hardness which resisted the imagination, which did battle in the name of fact against the images we live with'.

Tóibín shared then, I realised, my own resistance to science's dismissal of ordinary human experience (colour a con job!), but then seemed pleased with the idea of a sort of holy war against science. He writes:

There was, he felt, as he lay in bed in the darkened room of the hotel, a battle going on. That day he had moved among the foot soldiers of the other side; he had looked at their armaments and studied their plan for wars and invasions. By writing what he wrote in silence, by revising it as carefully as he could, he had begun a small but powerful insurgency; he had stirred up something that could not be easily contained as it flitted through the side streets of some city by night, finding shelter and a good vantage point in empty buildings. He had put a flag up for his own freedom, claiming precedence for the visible world, for the human presence, in all its foolishness, for the sweet cloud of unknowing.

As in all dualist positions, unknowing is prized as a badge of humanity. It's the same instinct that causes us to be so pleased when the scientists can't figure out what consciousness is. Still, it's hard to see what is intrinsically noble about unknowing; do we really want to yearn for the innocence of Eden? Wouldn't it be more *fruitful*, perhaps, to persuade science to be *scientific* about the imagination, to pluck the proverbial apple from the tree again and resolve once and for all the knowledge it constitutes? Do I intend the apple? Is the apple *represented* in my head by my neurons? Does it offer an affordance that I enact in grasping it? Is it a Bishop Berkeley apple of my mind? Or is it, the apple that is, experience itself? Which means *I* am not my body at all, but something my body creates when it enters into contact with the world? I am apple. And oranges and pears too of course.

All of a sudden, our *Wissenschaft* evening was over. Or rather the debate was. We could start drinking wine and eating quiche. It was after midnight. A hundred people crowded round a long table, then milled with plates and drinks. Hannah Monyer was there and introduced me to a quantum physics professor, Matthias Weidemüller, who said he held out high hopes that Giulio Tononi's integrated information would explain consciousness. When I asked him why on earth an intensification of neural interconnectedness, which was essentially what Tononi meant by his integrated information, should give rise to something as astonishing as consciousness, he said that quantum physics had shown that changes in some particular aspect of an environment could produce unpredictable changes and indeed 'emergences'. For example, nothing predicted that when you cooled water down, at a certain point it would become a solid. That was a remarkable metamorphosis.

This statement rather floored me. Water freezing to ice seemed a phenomenon of such an utterly different nature and entity than consciousness, and needless to say I felt ill equipped to take the matter on with a professor of quantum physics. Later, in my room, a little research would tell me that there was still much debate as to whether freezing was 'predictable' or not (in the sense that, if one didn't know water freezes would you be able to predict it from a study of the molecules?). In the meantime, however, I drained my wine, which was barely chilled, pleaded exhaustion and retreated to bed.

The following morning, I tackled Philosopher's Way again. There was no Eleonora this time, she had work to do at home. But thanks to her persuasion I did have a smartphone now and we exchanged messages from time to time as I crossed the bridge and tackled the hill. This also meant I could take photos now, as I hadn't done on the previous walk. And I could use the Net to look up translations of Joseph von Eichendorff's poem. So the objects of experience available to me, or simply the experiences, if you accept that object and experience are one, were considerably extended. With considerably more potential for distraction.

The weather was rather better than last time, bright with just a touch of mist on the water. Confident that I knew the way, I didn't bother with the hotel map, then found I wasn't altogether sure where I was supposed to make my right turn after the bridge. I had thought it was quite soon after, the first or second street, but actually it was a little further on. I had also thought that one turned directly into Philosophen-weg off the main street, whereas in fact you first had to do a block on Ladenburger Strasse. So, as with all returns, I remembered it and I didn't remember it. Eating breakfast at the hotel I had had a notion of how things would be, of what to expect. I had had an idea of the bridge, then a busy main street, the right turn, then the steep climb up a nar-row street which would gradually open into the countryside. I was aware, for example, that I would find the garden on the right with its memorials of Eichendorff and Hölderlin. I knew there would be stone walls and houses. But this was mainly semantic information. If I tried to *visualise* any-thing, to experience the walk in advance, as it were, it was all very vague. To a large extent this semantic memory, or re-construction rather, depended on building one thing on another, even deducing things. You went up the slope, so where did you come down? The path ran horizontally across the hill, so to the left there would be an ascent and to the right a descent, with occasional view of the river, and so on. I didn't really *see* any of this. And yet . . .

And yet as soon as I was on the path I knew I had been here before. A house with a lemony facade, the terracing in the garden – they were not new to me. Neural patterns laid down a year ago were presumably reactivated. That said, nothing was quite as I remembered it. And some houses I didn't remember at all, even though I *knew* as a fact that I had, if not seen them, at least passed them by. Perhaps un-seeing. So the experience of recognition was patchy. And this patchiness – recognition/non-recognition – changed the nature of this second visit to the Philosophenweg and made it radically different from the first visit. It was first visit plus a second. Or rather, a new visit, with bits of the old. In particular, the walk seemed much *shorter* than I remem-bered. I reached the famous garden in no time. Was this because I was walking faster without Eleonora? I don't think so. Or because without our conversation less had *hap-pened* on the walk? Who knows? In any event it was the same path, but its length now seemed quite different. At least in terms of time. It wasn't the same experience.

Nor did the image of Eichendorff in profile look quite as I remembered it. I had thought it was made of the same stone as the memorial itself, but actually, it was a dark green bronze sculpture riveted to the stone, so that though it gave the impression of a bas-relief it was actually a separate ob-ject, in sharp contrast to the pinkish stone behind. In any event, quite different from what I had remembered.

Why bother you with all this? These are typical observa-tions of any experience repeated a year on. Familiarity, unfamiliarity. Sameness, surprise. But I want to use this to try to say something more about what objects are, and how they stand in relation to us, or rather to our bodies.

In the standard vision of consciousness, the object is always there, much as we know it; the memorial to Eichen-dorff is always just inside the gate to the garden on

Philosophenweg. It stays put and it remains as we know it. Of course if we like we can try to imagine it as made up of rock crystals and minerals, or even of atoms, though at that point we may need to rely on diagrams seen in textbooks, rather than anything on the Philosophenweg. The object is always there, our brain has stored a memory of it, accurate or inaccurate as the case may be, an engram constructed of neural pathways and patterns. Perhaps in the hippocampus. Now when we are before the object again, the new experience is compared with the memory and the two integrated into a new experience which will also be a new memory. We don't doubt the absolute existence of the object.

But if we go to the Spread Mind theory, the object *is* the experience, hence it only happens when I am there. Of course similar experiences will happen when other people are there, and experiences of quite different kinds when an ant crawls up the face of the stone, or when a bird sits and sings on top of it, or shits, or when a dog comes to sniff the smell some other dog has deposited on it. In these cases the memorial to Joseph von Eichendorff happens in rather different ways. Or rather *part* of what I refer to as the memorial to Joseph von Eichendorff happens.

But what happens when I am not there, and no one else is there, and no animal is there? Then there is no experience of the memorial. And since experience and object are the same, and since, in this theory, there is no absolute object, but only the relative object created by its causal relation with whatever animal is experiencing it, what is there?

Gazing at the plaque to Eichendorff, I struggled with this idea, which has always been one of the most difficult ideas for me in the Spread Mind theory. And I couldn't accept that the memorial wasn't always there exactly as I saw it now, even when I wasn't there. Though of course it did look rather different from the memorial as I remembered it. I hadn't remembered that pink colour at all.

I read Eichendorff's poem again.

> *Schläft ein Lied in allen Dingen,*
> *Die da träumen fort und fort,*
> *Und die Welt hebt an zu singen,*
> *Triffst du nur das Zauberwort.*

Smiling, I took a photo of the stone with my phone and sent it via WhatsApp to Eleonora. Moments later she messaged back.

'Sleeps a song in all things, / Which do dream on and on, / And the world will start to sing, / If only you find the magic word.'

'You are my magic word,' I texted back and received a smiling emoticon in return.

Then spurred by I don't know what curiosity, I put Eichendorff's poem in Google and searched for a translation. First I tried Google Translate which, despite being much enhanced these days, still cannot know, poor thing, that it's translating, or indeed doing anything at all. It gave me:

> Sleep a song in all things,
> They are dreaming now and then,
> And the world is beginning to sing,
> You only meet the magic word.

Eleonora and I had done rather better than this, I thought. Google's program had read the first *Shläft* as an imperative! Imagine that. Not realising that you can't tell someone to sleep something (Johnny, sleep a song, will you!). Never mind. It's pretty amazing how well the computer does, given the handicap it starts from. And at least I now knew that the standard translation of *treffen* must be 'meet', since computer translations will always give you the most standard translation of a word if there is not some very obvious reason for doing otherwise. So it should be, 'If only you *meet* the magic word.' Meaning, if only you *come across it*, perhaps?

Now, still standing by the memorial, I typed 'Sleeps a song Eichendorff' into my phone and eventually found the following published translations:

> Sleeps a song in all things,
> That keep dreaming on and on,
> And the world will begin to sing,
> If only you find the magic word.

How close to our own that was! Only the 'keep' was different – keep dreaming – and even that was hardly a big change. But what about this:

> Sleeps a song in things abounding,
> that keep dreaming to be heard:
> Earth'es tunes will start resounding
> if you find the magic word.

Here, to keep the original's rhyme, three new notions have been introduced: 'abounding', which sounds very *Hymns Ancient & Modern*; then 'resounding', which confirms this feeling (at the expense of simple ideas like 'all things' and 'singing a song'); then, bringing in 'heard' to rhyme with 'word', we introduce the idea that our 'things abounding' are dreaming *to be heard*, the way a teenager might dream of stardom. 'Earth'es tunes' sound curiously tinny, while the 'only' of the last line, with its implication of how difficult the enterprise is, has gone.

But of course translations that are not done by computer programs are always different, because everyone hears the original a little differently, everyone has slightly different resources in the original language and a slightly different sense of what sounds right in their own language, not to mention a different understanding of the kind of faithfulness, or equivalence, one might be looking for in a translation. Even when we simply read a text we all respond differently. One reviewer is enthusiastic and another insulting. One student reads avidly and another tosses the book aside in disgust.

Everyone who stands in front of Eichendorff's memorial stone and reads the poem engraved there will have a slightly different impression. Or we could say, each of us brings the poem into being in different ways each time we come to it. No that's not quite right. If the experience, in this case the poem, is both subject and object, how can I say that 'we bring the poem into existence' since this would be tantamount to bringing oneself into existence? Let's say, rather, that reading a poem brings a 'we' or 'I' into existence, which, during these moments, is the poem. Either way, the poem is not the words on the page, but the experience that accompanies reading or reciting the words. So that actually the poem is different, or new, on each reading, even though the conditions that prompt all these 'poem experiences' remain exactly the same, which is to say, the signs on the page. Or engraved in stone. Translations are a demonstration of this, each one being an intersection of the original formula of words with the translator's sensibility and resources, at a particular moment, since one might well translate the poem differently on another occasion. And translating something a second time one is aware of the first time, which is very likely causally present as one begins again, perhaps drawing us back to the first version, or pushing us away from it. Or now one, now the other. So time is spread and the new translation is in relation to the old. Unless the old is completely forgotten.

Is this a useful analogy for perception itself?

Could one say that the monument to Eichendorff, and indeed the whole surrounding landscape, from the shrubs and trees of the terraced garden to the river far below and the trees climbing the slopes above, was, like the poem, waiting to *happen afresh* when I or anyone else, or rather my body or anyone else's, returned to see it? That it – the memorial stone – really is a different object from the object of a year ago, a different experience, a different 'I' brought into being, even though the conditions for my having that experience remain substantially the same?

I still find this very difficult. I know that when I leave the scene, or simply turn my back, the experience will be gone. But I find it so hard not to think that the memorial is there in my absence exactly as it is in my presence. And similarly available to the people who visit it every day.

This, along with his account of dreams, has always been a stumbling block for me with Riccardo's theory.

So for a while, the morning after the *Wissenschaft* conference, I stood looking at the monument to Eichendorff trying to think of its pinkish stone and bronze sculpted head as the *conditions* for experiencing what I experienced, conditions that required a relationship with a perceptive system, *my* perceptive system, to be the object it was for me, rather than an absolute object in its own right. It was hard work, and why I had fixed on just this stone to try to settle this issue once and for all I have no idea.

I looked at the stone, then looked at the photo of the stone on the phone. The photo I had sent to Eleonora. I took another photo from a slightly different angle and looked at that. One can take any number of photos of an object, or of the conditions that allow the photo to appear on the phone. On the photo I noticed there was a patch of moss on the stone, a yellowish patch perhaps six inches from the bronze sculpture following the sharp line of the nose downwards. I hadn't actually noticed this when looking at the stone itself. But turning from the photo to the stone, sure enough the patch was there. And for some reason I decided to touch it.

This was a turning point.

I touched the moss – or rather, again, my fingers touched the moss, producing an experience, 'I'. A soft-to-the-skin experience. My fingertips moved across the stone which was rough and grainy, with small blackish streaks running here and there from top right to bottom left. Of course my fingers didn't experience these black areas as black, or as streaks, but as very slight erosions, grooves. I pressed my hand flat on the stone. That was a different experience. I shut my eyes and grasped the side of the stone which was, and no doubt is, about three inches thick. Now the stone was a solid gritty thing. Without vision, the experience was quite different. This was a different object. And suddenly it was terribly easy to think that when I took my hand away, that object, that experience, would be gone. The experience/object, gritty feeling of cold damp hard surface, only existed when I touched the stone, but the conditions for my experiencing it again were still there, would go on being there even when I wasn't touching it.

So all one had to do was think of vision as touching and suddenly it was easy to appreciate that every time you moved your eyes the world happened anew, or rather a different object happened, from all the possible objects that can happen, and that object, produced by the meeting of those external conditions and my senses, was my consciousness, was me, or part of me. Simultaneously the world and my experience. We were the same thing.

No doubt there were a thousand complications to be taken into consideration. The body's perception of itself. The accumulation of previous but still present perception, conditioning the new. And language forever intruding as words from the past hurry to meet experiences in the present. All the same, this moment when I touched the moss on

Eichendorff's memorial, then shut my eyes and gripped the stone itself, was another major step towards taking on board the Spread Mind theory in all its strange obviousness.

'You don't need a magic word,' I texted Eleonora. 'You just need to be here.'

'Come again?' she responded.

'You are the song in *allen Dingen*.'

'A Wordsworth moment?'

'I'll explain,' I texted, 'when I get home and we are one.'

Afterword

This book is very much, and without apologies, the story of my reactions to the ideas of Riccardo Manzotti. 'But is he serious?' people ask when we discuss his Spread Mind hypothesis, 'does anyone else take his line?' Proudly liberal in our politics, always glad to repeat our commitment to freedom of speech and of thought, when it comes to metaphysics we crave authority, we struggle to shake off even the craziest ideas established wisdom upholds and hesitate so much as to consider those it does not. Fortunately, since I began work on this project, Manzotti's thinking has been finding a wider and wider audience. His account of afterimages – 'A Perception-Based Model of Complementary Afterimages' – was published in *Sage Journals* in January 2017. The academic publisher John Benjamins came out with *Consciousness and Object: A Mind-object Identity Physicalist Theory* in June and in November OR Books, New York, followed with *The Spread Mind: Why Consciousness and the World Are One.* A series of dialogues between Manzotti and myself on the online pages of *The New York Review of Books* (www.nybooks.com/topics/on-consciousness/) led to his being invited all over the world to explain his theory, not least, if somewhat bizarrely, alongside Deepak Chopra at the Science and Non-Duality Conference in San Jose, California. More importantly, for those anxious to hear of authoritative support, an article on hallucinations is shortly to appear written jointly with Alex Byrne, Professor of Philosophy at MIT and one of the world's foremost academics in this field.

That said, I would invite readers to refer everything they read about consciousness, whether received ideas or exciting new theories, to their own experience; never to be wowed or dazzled; scrupulously to consider what it's really like being alive. When it comes to consciousness, we are all repositories of quantities of evidence far richer than any available in the neuroscientist's laboratory.

On the other hand, let us not fall into the complacent scepticism that supposes nothing more can ever be learned about these matters. Why not? Very often the obstacle to knowledge is not some supposed technical barrier, but the collective mind-set, the huge interests inevitably invested in our present way of seeing the world. Why not go on exploring? Why not suppose that in some distant or perhaps not-so-distant future, people will wonder why we couldn't see what was, so to speak, right under our noses.

Tim Parks, October 2017

TIM PARKS has written eighteen novels, including *Europa*, which was short-listed for the Booker Prize, and most recently, *In Extremis*. He is the author of several works of nonfiction, including *Italian Neighbours*, *Italian Ways*, and *Where I'm Reading From: The Changing World of Books*. Parks has also translated the works of Alberto Moravia, Giacomo Leopardi, and Niccolò Machiavelli, among others, and is a frequent contributor to *The New York Review of Books* and the *London Review of Books*. He lives in Milan, Italy.